REPTILES *of the* NSW MURRAY CATCHMENT

A Guide to Their Identification, Ecology and Conservation

Damian Michael and David Lindenmayer

With contributions from Mason Crane, Matthew Herring
and Rebecca Montague-Drake

CSIRO
PUBLISHING

National Library of Australia Cataloguing-in-Publication entry

Michael, Damian.

Reptiles of the NSW Murray catchment: a guide to their identification, ecology and conservation/Damian Michael, David Lindenmayer.

9780643098206 (pbk.)

Includes index.
Bibliography.

Reptiles - Murray River Region (N.S.W-S. Aust) - Identification.
Habitat (Ecology) - Murray River Region (N.S.W-S. Aust).
Conservation biology - Murray River Region (N.S.W-S. Aust).

Lindenmayer, David.

597.909944

Published by
CSIRO PUBLISHING
36 Gardiner Road, Clayton VIC 3168
Private Bag 10, Clayton South VIC 3169
Australia

Telephone: [+613] 9545 8555
Local call: 1300 788 000 (Australia only)
Fax: +61 3 9662 7555
Email: csiropublishing@csiro.au
Web site: www.publishing.csiro.au

Front cover: photos by David Webb (top) and Damian Michael (bottom).

Back cover: photos by Damian Michael (left and right) and Peter Robertson (middle).

All photographs are by Damian Michael unless otherwise stated.

Set in 11/14 ITC Garamond
Edited and indexed by Lachlan Garland
Cover and text design by James Kelly
Typeset by Desktop Concepts Pty Ltd, Melbourne
Printed by Ingram Lightning Source

CSIRO PUBLISHING publishes and distributes scientific, technical and health science books and journals from Australia to a worldwide audience and conducts these activities autonomously from the research activities of the Commonwealth Scientific and Industrial Research Organisation (CSIRO).

Murray CMA CATCHMENT MANAGEMENT AUTHORITY

THE AUSTRALIAN NATIONAL UNIVERSITY

Feb26_RP_ILS

Contents

Preface

Many people realise Australia supports a diverse array of plants and animals, but few would be aware of just how species-rich particular groups of animals are. The number of Australian reptiles, for example, is unrivalled anywhere else in the world. The number of new species being described each year is increasing so rapidly that the actual number in Australia may exceed 1000. Some species are well known and even loathed by some, but most people would be surprised to learn just how many different kinds of reptiles are living in their neighbourhood. Few would be aware of large pythons living in rabbit burrows or 'worm-like' snakes tunnelling below the ground, or even how to tell a gecko from a skink. We have attempted to bridge this knowledge gap between ecologists and the general public by producing a field guide to the reptiles of the NSW Murray catchment. We acknowledge that farmers and landholders are the custodians of much of the land in the Murray catchment, and it is in their hands that the future of many Australian reptiles lies. We believe that a deeper understanding and appreciation of reptiles will be gained when people become familiar with this remarkable, and often misunderstood, group of animals. Only then will people take the time and make the effort to conserve these amazing creatures. Therefore, we hope this book stimulates interest in reptile conservation in Australia.

Damian Michael
David Lindenmayer

Acknowledgements

This book would not have been possible without the generous contributions made by numerous people. We particularly thank the New South Wales Murray Catchment Management Authority for financial assistance. The staff from the Murray CMA: David Leslie, Helen Wilson, Jack Chubb, Emmo Willinck and David Costello were extremely supportive of this book. It has also been a privilege to work with the CMA on long-term biodiversity monitoring programs such as the 'South West Slopes Restoration Study' and the 'Murray Biodiversity Baseline Monitoring Project'.

We are greatly indebted to all those who generously supplied photographs for this book. The contributors were: Peter Robertson, David Webb, Nick Clemann, David Hunter, Jeremy Tscharke, Hugh MacGregor, Lachlan McBurney, Matthew Herring, Craig Michael, Jerry Alexander, Steve Sass, David Blair, Clare McCutcheon, Dylan O'Donnell, Nick Edards and David De Angelis. We thank Dale Nimmo for assisting with sourcing some of the images and a sincere apology to all who supplied images that were not used.

We gratefully acknowledge the contribution made by the many volunteers and colleagues who enthusiastically helped with aspects of field work over the years – Lachlan McBurney, Chris McGregor, Rachel Muntz, Sachiko Ellicott, Steve Holliday, Scott Lucas, Greg Slade, Hugh MacGregor, Nigel Jones and Geoff Kay. We sincerely thank the landholders of the Murray catchment for supporting our research and allowing us access to their properties – the list of participating landholders is too many to mention.

DM would like to thank his wife, Tracy, who provided continuous support, encouragement and constructive criticism throughout the writing of this book. Clive Hilliker prepared the location and distribution maps, Tracy Michael prepared the vegetation map and Rachel Muntz did an excellent job co-ordinating many aspects of the work associated with this book. Finally, DM would like to thank his family for fostering an environment that allowed him to develop a passion for wildlife conservation.

1 Australia - the land of reptiles

Australia is often referred to as the land of parrots, but it could just as appropriately be called the land of reptiles. This is because the number of Australian snakes and lizards far exceeds the number of bird species. In fact, Australia has more reptile species than any other country in the world. We have three times more reptiles than the United States of America. There are many other places in the world that support high numbers of reptiles. These places are often called 'biodiversity hotspots'. Many islands, such as Madagascar, Sri Lanka and those of the Caribbean, are included in the list of global hotspots. However, as far as continental islands are concerned, no other country comes close to matching Australia.

Exactly how many different kinds of reptile species call Australia home is currently unknown. By late 2008, over 880 reptile species had been formally described. However, the discovery of new species is continuing at such an astounding rate that, since the early 1980s, more than a dozen new species have been described each year. This trend has lead scientists to predict the actual number of reptile species in Australia may soon exceed 1000. The rapid growth of Australia's identified reptile species is testament to the increased interest in the field of herpetology, as well as an increase in genetic research. Generally, new species are described when an existing species is split into separate species based on molecular and other genetic techniques. For example, 13 new skinks in the genus *Cryptoblepharus* and 16 new geckos in the genus *Diplodactylus* were discovered in 2008 and 2009 using these methods. However, the discovery of new species is not always a result of genetic research. In 2006, scientists from the Museum of South Australia literally stumbled upon a large 'brown-looking' snake in central Australia. This snake turned out to be a new (and third) species of taipan (one of Australia's most venomous snake species).

It is often exciting when new species are discovered because it shows how little humans known about the wonders of the natural world. Australia is no exception, as there is much to be learnt about the different reptile species that inhabit the continent. In many cases, a basic understanding of where a particular species' occurs, what habitats they live in and what they prefer to eat are not known. Other

Steve Sass

One hectare of spinifex-dominated woodland can support more species of reptiles than any other vegetation type in the world of comparable size.

information such as how long reptiles live and how many offspring they have is also lacking. Not surprisingly, many reptiles don't yet have a formal common name. For anyone wishing to enter a branch of science that is full of mystery, the field of herpetology may be for you.

Australia's remarkable reptile diversity

Why Australia supports more reptile species than other continents is a subject of much scientific debate. Some scientists believe it may be due to our long period of stability, isolation and historical climate patterns favourable to the evolution of different reptile species. Reptiles that cope well with stable conditions have had a long time to exploit a wide range of different ecological niches. In contrast, in times of prolonged difficult climatic conditions, many species became specialised to deal with localised climate regimes. Other scientists have suggested that high reptile diversity may be due to the myriad of different habitats found in Australia. This includes the widespread occurrence of spinifex grasslands, which dominate arid regions and support more reptile species than comparable areas of any other vegetation type in the world. Often, when spinifex grasslands are burnt, they create a mosaic of different aged vegetation. Over time, this allows a greater number of species to live in the one area. The abundance and diversity of nutrient-rich food sources such as termites is another proposed reason to explain Australia's high reptile diversity.

Reptile facts and myths

Almost any Australian who has spent time on the land has a snake story. This has been incorporated into Australian folklore and is reflected by the number of snake myths. Some common misconceptions are that snakes are slimy to touch, death adders have a sting in their tail, snakes are attracted to milk, the only good snake is a dead snake, and hoop snakes bite their tail, form a loop, and roll away when threatened. There is little truth to any of these myths but other reptile facts should be made known. Many people are aware that Australia has more venomous than non-venomous snakes, although few know that only a dozen of the venomous species have the potential to cause serious harm. It is human nature to be fascinated by the few animals that are dangerous. But the 868 other known Australian reptiles are just as worthy of our attention and fascination. For example, over 90% of Australia's reptiles are found nowhere else in the world and over half of the world's pythons and goannas live in Australia. We have one of the world's longest snakes, the Australian scrub python, which can exceed 6 metres in length; the world's smallest python, the pygmy python which averages less than 60 centimetres, and

Craig Michael

The thorny devil (*Moloch horridus*) is an iconic Australian lizard that lives in arid parts of central Australia and feeds principally on small black ants.

the world's smallest goanna, the short-tailed pygmy monitor which grows to just over 20 centimetres. Some snakes even feed on animals not usually associated with reptiles. For example, the white-bellied mangrove snake feeds almost entirely on crabs, and blind snakes eat the eggs and larvae of ants. The thorny devil also makes a living on ants. It can consume over 5000 ants in a single day and remarkably, drinks water by simply standing in moist sand. Water travels up the thorny devils legs to the corners of its mouth via capillary action along small grooves between the scales. Despite these and numerous other interesting facts, for some people, reptiles will always be feared and loathed, but for those that are not squeamish there is no better place than Australia to appreciate reptilian fauna.

Aims of the book

This book aims to help landholders, natural resource managers and interested members of the general public identify any species of reptile encountered within the NSW Murray catchment. Identifying reptiles in the field can be difficult. Therefore, we hope by producing a regional guide, readers will easily be able to identify any of the 80 reptile species found in the NSW Murray catchment. Being able to distinguish between different species is an important step in conserving them. A vast number of reptiles occur on private property. Hence, there is growing need to manage reptile habitat in farming landscapes, while at the same time maintaining productive and sustainable farming enterprises. This book provides practical advice to landholders and natural resource managers who wish to learn more about integrating reptile conservation and farm production.

Scope of the book

In Chapter 2, we explain why it is important to conserve reptiles, their role in the food chain, their use in human medical research and the main threats to their survival. Chapter 3 provides practical advice on how to manage reptile habitat in farming landscapes, and Chapter 4 describes the NSW Murray catchment area, including a summary of the broad vegetation types and plant species found in the region. Species accounts and definitions used in this book are provided in Chapter 5. This chapter includes images and descriptions of all 80 reptile species found in the Murray catchment, as well as information on key distinguishing features, distribution and preferred vegetation types. Each species' account includes information on adult body length, life-form, reproduction, activity, commonly used shelter-sites and conservation status in the region. The distribution maps for each species are based on records contained in the wildlife atlas of the Department of the Environment and Climate Change (formally the NSW National Parks and Wildlife) and the Museum of Australia, as well as extensive surveys carried out by wildlife ecologists from The Australian National University and Murray Wildlife Pty Ltd. We also include additional natural history information, a section on interesting facts and a list of management topics for each species. In Chapter 6, we provide summary tables of the main distinguishing features that will help readers quickly identify similar looking skink species in the Murray catchment. In Appendix 1, we provide a checklist of species and the main vegetation types they are associated with in the region. A list of all other species of animals and plants mentioned in this book is provided in Appendix 2.

2 The importance of conserving reptile diversity

There are several reasons why conserving reptiles is important. First, they are part of Australia's unique natural and cultural heritage. Future generations should be able to experience and enjoy these animals in their natural environment and not just in wildlife parks and zoos. Some species are far less common than they used to be. It would be a great shame if future generations were to only learn about reptiles from pictures in books or from stories passed down by their grandparents. This chapter explores some of the other reasons why conserving reptiles in Australia is important.

Reptiles in the food chain

Reptiles are an important food source for a wide range of animals, including birds and small native marsupials such as the yellow-footed antechinus and the fat-tailed dunnart. Birds such as the laughing kookaburra, pied butcherbird and the bush stone-curlew eat reptiles. Raptors such as the black kite and brown falcon prey on snakes and the white-bellied sea-eagle has even been known to capture large pythons. Lizards are an important part of the diet of other reptiles. Goannas move with lightening speed and will pursue and capture almost any reptile they encounter. They will even tackle one of Australia's largest venomous snakes, the mulga snake (which is often referred to as the king brown snake). Many snakes and some of the larger lizards prey on skinks and geckos and some reptiles feed exclusively on lizards. These species are called 'saurophagus' reptiles. Similarly, some snake species feed entirely on other snakes, even their own kind. These species are called 'ophiophagus' reptiles.

Some species of reptiles are top predators in the food chain. Top predators play an important role in any ecosystem because they can help regulate populations of animals further down the food chain. For example, brown snakes play an important role in controlling introduced mice, and carpet pythons, in some parts of Australia feed exclusively on the introduced black rat and European rabbit. When these

Many skinks, such as the woodland tussock skink (*Pseudemoia entrecasteauxii*), feed on a broad range of invertebrates ensuring they play an important role in controlling insects in agricultural landscapes.

kinds of reptiles are killed or disappear from parts of the landscape, pest species may multiply with no predators to control numbers. Similarly, goannas will feed on carrion, a service that helps to prevent diseases from spreading in agricultural areas. However, the majority of lizards, such as geckos and skinks feed on insects and other small invertebrates. These species may help control crop pests and effectively provide landholders with a free environmental service.

Reptiles in human medical research

Some species of reptiles have the potential to provide new insights into human medical research. Preliminary studies show snake venom can slow the growth rate of cancers. Other compounds found in snake venom could be used to develop blood clotting skin sprays, which may help prevent bleeding from severe wounds. The venom from brown snakes is being used as a diagnostic tool in testing for the debilitating disease lupus and tiger snake venom is being used to better understand how the human neuromuscular system works. Snake toxins are complex compounds and can vary between species as well as between populations of the same species. Therefore, in the interests of medical research, populations of as many different species should be conserved, because many potentially useful substances may be waiting to be discovered.

Reptiles and amphibians, particularly the salamander and some frog species, have the ability to regenerate their tails or limbs. Pioneering research on American salamanders is providing useful insights into limb regeneration, which could, one day, be applied to humans. This may come sooner than expected as scientists in Japan have recently discovered a molecule in a species of frog which stimulates the regenerative process and may be a key to understanding limb regeneration in other vertebrates such as reptiles.

Threats to reptiles

There have been unprecedented declines to many Australian native animals, including reptiles. While no known species of reptile has become extinct since European settlement, many species are much less abundant than they used to be. This is because of threatening processes such as loss of habitat, altered fire regimes, changed hydrological regimes, over-grazing by livestock, broad-scale cultivation of crops, weed infestation, pesticide use, predation by the red fox and feral cats, and soil erosion caused by rabbits. Conservation reserves can ameliorate some of these threatening processes. However, the majority of reptile species in the Murray catchment occur predominantly on privately owned land. Therefore, relying solely on nature reserves will not result in effective conservation. This is because conservation reserves do not adequately encompass all vegetation types, habitats and reptile species in the region. Landholders and natural resource management agencies also are often in positions to make relatively minor modifications to farming practices which may benefit a wide variety of species. Financial incentives and stewardship payments are available to landholders who wish to manage parts of their property for wildlife conservation. For further information contact your local Catchment Management Authority or Landcare Group.

Damian Michael

Grassy box woodland dominated by white box, yellow box and Blakely's red gum occurs on productive soils and is listed as an endangered ecological community in New South Wales.

Reptiles also may be susceptible to the effects of climate change. For example, in mountain environments, some species of reptile are being forced to move upwards to compensate for the loss of habitat at lower elevations. These species are effectively trapped because they have nowhere to migrate. Reptiles with offspring that are determined by the temperature of the nest (e.g. turtles, dragons and many skinks) may be unable to evolve quickly enough to counteract the consequences of increasing temperatures. In the Murray catchment, drought and altered flooding regimes is impacting on the health of river red gum forests. In some areas, grey box trees are encroaching onto the floodplain and replacing river red gum trees. This may have severe consequences for species that rely on river red gum forest environments. The effect of climate change also may interact with existing threatening process like altered fire regimes, and invasive plant and animal species. In the next chapter we explore the key areas of management that are required to help conserve reptiles in farming landscapes.

3 Conserving reptiles on farms

Many people wish they could do something to help reverse the decline of Australian native wildlife. This chapter provides some practical advice on how reptile habitats might be best managed in farming landscapes. This section has been written primarily with landholders and natural resource managers in mind. However, many of the basic principles discussed can be adopted in urban back gardens, hobby farms and conservation reserves.

Management themes and actions

Soil structure and integrity

Intact and uncultivated areas of bare ground with a lichen-encrusted surface, soil cracks and invertebrate tunnels are used for basking and foraging by dragons, skinks and geckos, and for sheltering by soil-dwelling lizards and blind snakes. Species such as the hooded scaly-foot and geckos use spider tunnels as shelter.

Damian Michael

Management actions: Once intact soils have been disturbed they often take a long time to recover, sometimes more than 50 years. Where possible, minimise the amount of soil disturbance when cultivating paddocks. Try to minimise the use of fertilisers, as they promote the growth of invasive plant species and change soil nutrient levels. Maintain low stocking rates or fence off areas with high native plant diversity, lichen-encrusted soil and invertebrate tunnels.

Patches of native grass

Areas of high quality native perennial and annual native grass (including spinifex-dominated mallee woodland) are used for foraging and sheltering by skinks, legless lizards and elapid snakes. Legless lizards, in particular, shelter within the base of tussock clumps.

Damian Michael

Management actions: Maintain low stocking rates and avoid over-grazing, fertilising or cultivating these areas. This is because once native plant species are replaced by exotic plants it can be a time consuming and expensive process to remove them. Native pastures are adapted to drought conditions and respond to periodic rain events better than exotic species. Native grass revegetation is possible by collecting seeds and spreading them over areas that have had exotic plant species suppressed (e.g. by spraying). In many cases, native grass species will naturally regenerate in response to ecological burning.

Rocky outcrops

Rocky outcrops, in particular granite, are used for foraging, sheltering, basking and nesting by a wide variety of species. In the South West Slopes bioregion within the Murray catchment, some declining species, such as the carpet python, are found only on granite outcrops.

Damian Michael

Management actions: In many cases, fencing off granite outcrops from livestock and implementing an appropriate grazing strategy is necessary to rehabilitate degraded rocky outcrops. Pest animals should be controlled to minimise the effect of predation. Invasive plants may need to be controlled to prevent over-shading of basking sites. Where possible, plant trees to link rocky outcrops with other landscape features such as creek lines and remnant vegetation. One important factor in outcrop management is to not allow rocks to become shaded by regenerating or planted vegetation. In many cases, periodic grazing and low intensity ecological burning may be required to control dense regrowth and invasive plants.

Wetlands

Wetlands and waterways are used as foraging and sheltering sites by turtles, some dragons, snakes and skinks. Access to water is important, as many species such as large snakes need to drink at regular intervals.

Damian Michael

Management actions: Fencing off wetlands and waterways from livestock will prevent over-grazing, soil erosion and excess nutrients from entering waterways. Where possible, allow natural flooding regimes to occur by preventing wetlands and billabongs from becoming disconnected from major waterways. In some cases, silted drainage lines may need to be excavated to allow this to happen. Water quality can be improved by planting native species to act as natural filtration systems. Basking sites for reptiles can be preserved by retaining snags and trees that fall into waterways. Apply caution when driving near wetlands after summer rains as this is a time when turtles actively cross roads. Remove illegal fishing nets found in river systems.

Tree plantings

Tree plantings are used as foraging and sheltering sites by skinks, geckos, legless lizards and some snakes. Some species will colonise tree plantings quicker than others, depending on the species' mobility, dispersal behaviour and condition of the surrounding landscape.

Damian Michael

Management actions: Tree plantings can be an effective way of creating habitat for some species of reptiles. Establish large block-shaped tree plantings with a mixture of tree and shrub species that are widely spaced (approximately every 2 metres). This will allow enough sunlight to penetrate the canopy and reach the ground layer where most reptiles forage and bask. Avoid high-density plantings in areas with abundant surface rock habitat or around granite outcrops, as they can contribute to negative shading effects. Include tree species that produce abundant stripes of bark to create habitat for arboreal reptiles. Where possible, limit grazing of tree plantings unless invasive plants are a problem. Leave dead trees and shrubs in plantings because they create extra habitat.

Leaf litter and fine woody debris

Leaf litter and fine woody debris that accumulate beneath shrubs and trees are used for foraging, shelter and nesting sites by skinks, geckos, legless lizards, pythons and snakes. Invertebrates that live in the leaf litter provide many reptiles with an abundant food source. Reptiles that incubate eggs, such as pythons, require dense mats of litter to construct nesting chambers.

Damian Michael

Management actions: Where possible, avoid raking, piling and burning leaf litter beneath paddock trees or in remnant patches of native vegetation. If fire is being used as a conservation tool, avoid widespread or complete 'blackout' burns. Instead, apply a fire regime that leaves some areas of native vegetation unburnt.

Surface rocks

Surface rocks are used for sheltering, basking and nesting by a wide variety of species from all reptile families. The thermal properties of surface rocks enable nocturnal species such as geckos to absorb enough heat to be active into the night.

Damian Michael

Management actions: Avoid removing bush rock from grazing areas and, if necessary, relocate surface rocks to other parts of a property instead of piling them up or selling them. Surface rock habitat cannot readily be replaced once lost, although substrates such as concrete slabs and terracotta roof tiles are an effective way of helping to artificially restore and enhance rock habitat.

Fallen timber and dead trees

Fallen timber and dead standing trees are used for basking, foraging, sheltering and nesting by a wide range of species. Reptiles that bask in direct sunlight often use dead timber because of the better thermal properties than living timber.

Damian Michael

Management actions: Where possible, retain fallen timber, dead trees and tree stumps in grazing areas and avoid stockpiling or burning them. Excess timber in one area can be redistributed to create habitat corridors around paddock boundaries or placed in tree plantings. Fallen timber habitats can be enhanced by introducing artificial shelters such as disused fence posts, timber off-cuts and railway sleepers in areas devoid of fallen timber. It is important to note that dead standing trees will eventually become fallen timber – a resource widely used by many different reptile species.

Shrubs

Shrubs are used for foraging, basking, sheltering and nesting by skinks, geckos, legless lizards and snakes. In areas naturally devoid of surface rocks or fallen timber such as chenopod shrubland, low-growing vegetation is a critical component of their habitat.

Damian Michael

Management actions: Retain areas with high shrub diversity and minimise grazing pressure to allow natural regeneration to occur. In some cases, direct sowing of seed in remnant vegetation will be necessary to restore shrub habitat. It is important to allow shrubs to die naturally and break down as they provide additional habitat in the form of fine woody debris and contribute to improved soil conditions.

Large mature trees

Mature trees, whether they are paddock trees or they occur within patches of remnant vegetation, are used for foraging, basking, sheltering and nesting by geckos, goannas, pythons and some skinks. Large trees often contain hollows and cavities that are critical in the survival of the carpet python and lace monitor.

Damian Michael

Management actions: Protect paddock trees by fencing them off from livestock to allow natural regeneration to occur. Develop a planting program, whereby new trees are planted to replace trees that may be lost during unplanned natural events such as fire and drought. Avoid using pesticides and fertilisers and prevent livestock from camping around the base of mature trees as these factors can contribute to reduced tree health and dieback.

Invasive exotic plants

Woody weeds and other invasive plants can shade basking sites, block rocky shelter-sites and reduce foraging areas for many species of reptile.

Damian Michael

Management actions: Woody weeds can be killed by cutting the stems and applying herbicide. Other invasive plants can be controlled by periodic heavy grazing or grazing during the flowering period of targeted species. Minimising the amount of fertiliser will benefit native plant species and allow them to compete successfully with exotic annual plant species.

Exotic animals

Some exotic animals such as the red fox and the feral cat can have a major impact on reptiles by preying on eggs, young and even adults. Exotic herbivores contribute to grazing pressure, soil erosion and soil nitrification, and in turn reduce habitat suitability for reptiles.

Management actions: A co-ordinated fox-baiting program across several adjoining farms can help control populations of the red fox. The feral cat can prey heavily on small nocturnal species such as geckos, legless lizards, blind snakes and elapid snakes. Trapping and shooting can be effective methods in controlling the feral cat, while domestic cats should be kept indoors at night. The European rabbit can cause extensive soil erosion and contribute to grazing pressure. However, their control has to be carefully weighed against the conservation of declining species such as goannas and pythons, which in some areas, feed primarily on rabbits or use rabbit burrows for shelter. Mice and rats pose little threat to reptiles, and in the case of brown snake, may have even benefited this species. Controlling rodents in and around the home will minimise the chance of encountering highly venomous snakes.

4 The NSW Murray catchment

This book focuses on the 35 500 sq km Murray Catchment Management Area (CMA) of New South Wales (hereafter called the Murray catchment). The area is bordered by the Australian Alps in the east (incorporating westerly flowing creeks and drainage systems), the Murray River in the south, the Murrumbidgee River water catchment divide in the north, and the intersection of both these rivers in the west (Figure 1). Major towns include Albury, Corowa, Deniliquin, Finley, Holbrook, Jerilderie,

Figure 1 Location of the Murray Catchment Management Area in New South Wales. (Source: DECC)

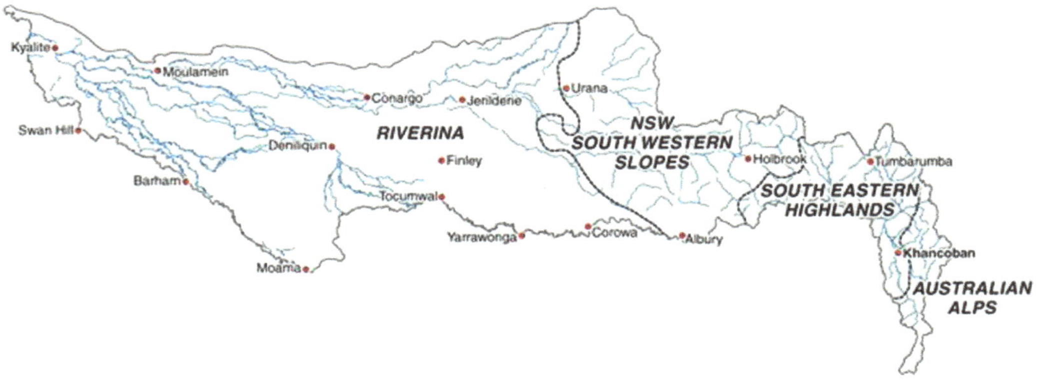

Figure 2 Major towns, waterways and bioregions of the NSW Murray Catchment Management Area. (Source: DECC)

Tumbarumba and Moulamein. Significant waterways include the Murray, Edward, Wakool and Niemur river systems and the Billabong Creek (Figure 2). Agriculture is the dominant land use and covers approximately three-quarters of the region and includes dryland cropping, livestock grazing and irrigation. Other industries include forestry and tourism. Protected areas such as National Parks occur in the east and travelling stock reserves and State Forests occur in the central and western parts of the catchment.

Climate

The climate of the Murray catchment is predominantly described as cool-temperate. January and February are the hottest months and the average minimum and maximum summer temperature ranges from 16°C to 32°C in the west and 6°C to 21°C in the east. July is the coldest month and the average minimum and maximum winter temperature ranges from 4°C to 14°C in the west and –4°C to 4°C in the east. Temperatures can be quite variable from year to year and droughts are common. The catchment spans over 2000 metres in elevation and experiences one of the largest ranges in rainfall of any catchment in Australia (320 mm in the west to 1600 mm in the east). At the highest point, Mount Kosciuszko (which in located in the Alps), is 2228 metres above sea level and is the highest mountain on mainland Australia. In the west, the lowest point is just over 70 metres above sea level near the junction of the Murray and Murrumbidgee rivers.

Reptiles in the Murray catchment

Until recently only a limited number of reptile surveys had been conducted in the Murray catchment. This was evident by the paucity of records in the wildlife atlas databases of the Department of Environment and Climate Change (DECC) and the Museum of Australia. However, over the past 10 years research on reptiles in the catchment has recorded more than a dozen species previously not known to occur in the region.

The Murray catchment supports 80 species of reptiles from nine families. These include: three turtles, nine geckos, four legless lizards, three goannas, five dragons, 38 skinks, four blind snakes, one python and 13 elapid snakes. Some species will be subject to new or further genetic studies which may reveal additional new species. Approximately one-quarter of the 80 species have an extremely restricted distribution. By 'restricted' we mean the species is known from only one or two locations. For example, beaked, beaded, gibber and thick-tailed geckos are found only in the far western part of the region. Likewise, the alpine she-oak skink and Snowy Mountains rock skink are found only in far eastern part of the region. In contrast, some species such as the eastern brown snake, large striped skink, southern marbled gecko and Boulenger's skink are widely distributed and found throughout the catchment.

It is likely that future surveys will detect additional species that have distributions close to the boundary of the Murray catchment. Some examples of these species include, the alpine bog skink and the weasel skink in the east, the striped legless lizard and the little whip snake in the central east, and the obscure skink in the far west.

A broad range in elevation and rainfall is the main reason why the Murray catchment supports many kinds of vegetation communities. These broad vegetation types, in turn, provide a variety of habitats which different species of reptiles can

The weasel skink (*Saproscincus mustelinus*) is known from north-east Victoria but is yet to be recorded in the NSW Murray catchment.

Damian Michael

The little whip snake (*Parasuta flagellum*) is known from the Murrumbidgee catchment in NSW and north-eastern Victoria but is yet to be recorded in the NSW Murray catchment.

exploit. Some species are adapted to the cool mountainous climates of the Alps, whereas others are adapted to living in hot, arid conditions. The South West Slopes (SWS) bioregion, which covers the area between the foothills of the Great Dividing Range (South Eastern Highlands bioregion) and the beginning of the Riverina plains (Riverina bioregion), is a transitional zone in which arid reptile species are replaced with cool-temperate species. The relevance of these broad vegetation types to the distribution of reptiles is discussed in the next section, with a summary of the type of plant species that characterise each vegetation type in the Murray catchment.

Vegetation type is vital to where particular reptiles occur. This is because vegetation and vegetation-related resources such as leaf litter, peeling bark and fallen timber are important habitats for most species. Furthermore, vegetation types vary in tree density and number of vegetation layers. These attributes influence the amount of sunlight that is able to penetrate the canopy and reach the ground - the place where most reptiles live. Shady and sunny areas are important because reptiles rely on external heat sources to maintain and regulate their body temperature. Some species are therefore better adapted to living in some vegetation types than others. For example, only those species which function at relatively low body temperatures can survive in heavily shaded, wet mountainous forests.

Many of the vegetation types, particularly in the SWS and Riverina bioregions, have been extensively cleared for agriculture and replaced with exotic grass species and crops. Exactly how extensive each vegetation type was prior to European settlement is often difficult to determine. However, an understanding of past vegetation patterns is important in understanding current distribution patterns of reptiles.

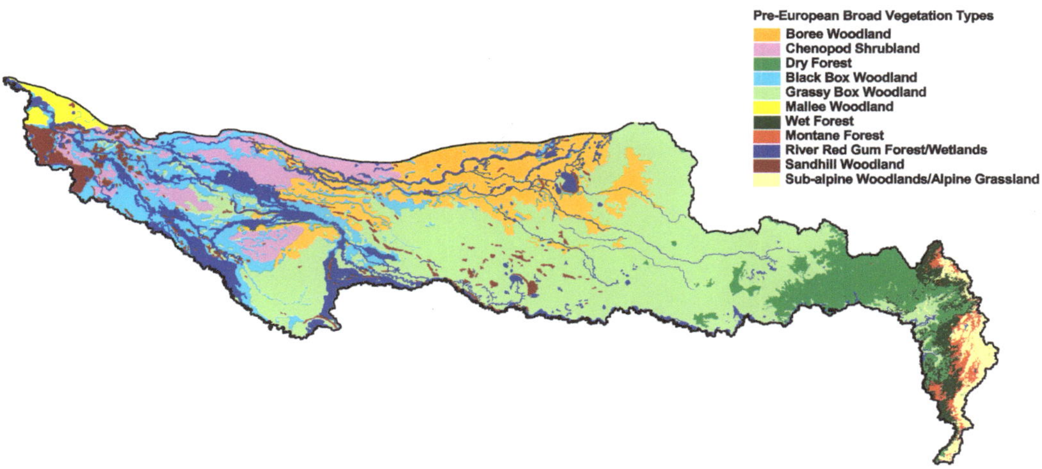

Figure 3 Map showing the extent of broad vegetation types in the NSW Murray catchment prior to European settlement. Note: alpine grassland is combined with subalpine woodland and wetlands are combined with river red gum forest. (Source: adapted from DLWC 2001)

Vegetation types of the NSW Murray catchment: from the mountains to the mallee

The NSW Murray catchment contains 13 broad vegetation types (Figure 3). The classification of these vegetation types is based on the dominant overstorey plant species (with the exception of wetlands and grasslands). Often, the actual mix of species will vary from area to area, giving rise to many different plant communities within the same broad vegetation type. The vegetation types in the Murray catchment change in relation to elevation, aspect, rainfall, topography and soil conditions. In alpine areas, broad vegetation types consist of treeless plains or low, stunted woodlands and support herb-rich, grassy ground layers. Slightly west of these alpine areas, rainfall and temperature increases, allowing taller trees and shrubs to grow. At the base of the Great Dividing Range, temperatures increase and rainfall begins to decrease, causing the forests to give way to the open grassy woodlands that typify the SWS bioregion. These undulating hills eventually merge with the plains of the Riverina, a region which supports a number of woodland vegetation types that are influenced by flooding regimes and soil type.

The plant species mentioned in the following sections are based on Costermans (1983) and Keith (2004) and are a broad indication of the types of plant species that occur in each broad vegetation type. These species may provide a useful guide in selecting plants in future restoration and tree planting programs. Trees and shrubs will often survive and grow better in places they are pre-adapted to, and native wildlife, including reptiles, will benefit from planted vegetation that is indigenous to a particular area. For additional information refer to the revegetation guides by Driver and Porteners (1993) and Kent *et al.* (2002) for the Riverina bioregion and Stelling (1998) for the SWS bioregion.

Alpine Grassland. Alpine grasslands and herbfields occur in the far eastern part of the catchment above 1500 metres elevation and are found in Kosciuszko National Park. The ground cover is often dominated by a rich variety of grasses, heaths and daisy species. Granite outcrops are a dominant part of the landscape and snow cover can last for up to four months each year. In wet depressions, sphagnum bogs occur.

Damian Michael

Subalpine Woodland. Subalpine woodland occurs in the far eastern part of the catchment above 1000 metres elevation and is predominantly confined to Kosciuszko National Park. The overstorey is often dominated by snow gum, although black sallee occurs in cold air drainage lines. The understorey consists of low shrubs including silver wattle, leafy bossiaea, daphne heath and narrow-leaf bitter-pea. The ground cover is often dominated by tussock grass, kangaroo grass, herbaceous plants and creepers.

Damian Michael

Montane Forest. Montane forests are found in an elevation band from 800 to 1400 metres. Overstorey species are tall and include pure stands of alpine ash, although mountain gum and snow gum may also occur. Understorey species include narrow-leaf bitter-pea, mountain mirbelia, broad-leaved bitter-pea and forest wattle. The ground cover is often dominated by wheat grass, blue flax-lily, bidgee-widgee and mother shield-fern.

Damian Michael

Wet Forest. Wet sclerophyll forests occur in the eastern part of the catchment in areas that exceed 1000 mm of rainfall per year. Overstorey species are typically tall and include narrow-leaved peppermint, eurabbie and ribbon gum. Damp depressions support swamp gum and mountain swamp gum. The understorey is dominated by silver wattle, blackwood, blackthorn, hazel pomaderris and musk daisy-bush. The ground cover is often dominated by tussock sedge, necklace fern and low growing herbaceous plants.

Damian Michael

Dry Forest. Dry sclerophyll or 'foothill forests' occur in the eastern part of the catchment between 300 and 700 metres elevation and in areas such as Woomargama National Park, Jingellic Nature Reserve and Bogandyra National Park. Overstorey species include red stringybark, broad-leaved peppermint, white gum, brittle gum, long-leaved box, apple box and red box. Black cypress-pine usually grows near rocky areas. The understorey supports a diverse range of shrubs, including Deane's wattle, nodding blue-lily, hop bitter-pea, handsome flat-pea, dogwood and grey guinea-flower. The ground cover is often dominated by wallaby grass, wattle mat-rush and rock fern.

Grassy Box Woodland. Grassy box woodlands occur on the fertile soils in the central part of the catchment usually below 700 metres elevation. Overstorey species include white box, yellow box, Blakely's red gum and red box. Grey box and white cypress pine become more common on the Riverine plains, while kurrajong, dropping sheoak and currawang grow on rocky outcrops in the east. Understorey species include low growing leguminous shrubs such as golden wattle, western silver wattle, montane wattle, gold-dust wattle and kangaroo thorn. The ground cover is often dominated by wire grass, white-top wallaby grass, plains grass, kangaroo grass as well as a diverse range of herbaceous, annual wildflowers. Valley floors and slopes are often grassy and devoid of shrubs.

Sandhill Woodland. Sandhill woodlands occur on alluvial sand dunes and prior streams that occur in the central to western part of the catchment. Overstorey species include white cypress pine, yellow box, buloke, Murray pine, belah and large shrubs such as rosewood, needlewood, butter bush, sweet quandong and yarran. The understorey includes emu bush, water bush and hop bush. The ground cover is often dominated by wire grass, curly windmill grass, spear grass and low growing herbaceous plants such as purple burr-daisy and ridge sida.

Matthew Herring

Black Box Woodland. Black box woodlands occur in the western part of the catchment on heavy self-mulching clay soils and on higher parts of the floodplain. Overstorey species include black box, although river red gum occurs frequently in flooded areas, and grey box and yellow box may be present in eastern ranging black box woodlands. The understorey includes thorny saltbush, slender-fruit saltbush, creeping saltbush, leafless cherry, lignum and nitre goosefoot. The ground cover is sometimes dominated by knob sedge, blown grass, panic grass and climbing saltbush.

Damian Michael

River Red Gum Forest. River red gum forests occur on the low-lying areas of the floodplain, along waterways and wetlands and are dominated by river red gum. They usually lack shrubs, but when present, understorey species include silver wattle, river cooba, cooba and dwarf cherry. The ground cover is often dominated by common reed, giant rush, river tussock, tussock sedge and water couch as well as a range of herbaceous plants such as lesser joyweed, common sneezeweed, swamp dock and river bluebell.

Matthew Herring

Boree Woodland. Boree woodlands occur on grey clay soils in the central northern part of the catchment and often intergrade with black box. Overstorey species include boree, which are often parasitised by grey mistletoe. Other less dominant species, such as rosewood and miljee may also be present. The understorey includes thorny saltbush, black roly-poly and black cotton-bush. The ground cover is often dominated by plains grass, rough spear grass, small-flowered wallaby grass, windmill grass and herbaceous plants such as rough burr-daisy, Australian bindweed and quena.

Matthew Herring

Chenopod Shrubland. Chenopod shrubland occurs on grey-brown, salty soils in the north-western part of the catchment and often intergrade with boree and black box woodlands. It consists of almost treeless plains dominated by a diverse range of saltbush species, including black bluebush, dillon bush, cotton bush, old man saltbush and bladder saltbush. These plants may be interspersed with annual grasses and a wide variety of herbaceous plants such as white-top wallaby grass, windmill grass, fairy grass, native carrot, hard-head daisy and pale beauty-heads.

Damian Michael

Mallee Woodland. Mallee woodland occurs in the far western part of the catchment on wind swept, sandy soils. Overstorey species include pointed mallee, ridge-fruited mallee, red mallee and weeping pittosporum. The understorey consists of low growing shrubs including grey wattle, comb spider flower, showy daisy-bush and wedge-leaf hop-bush. The ground cover is often dominated by small purslane, cactus pea, pale twinleaf, creeping saltbush and grasses such as rough spear grass and spinifex (porcupine grass).

Steve Sass

Wetlands. Wetlands occur throughout the catchment and support a range of different vegetation communities including river red gum forest and black box woodland. They support a wide variety of aquatic plants depending on their location, source of water supply and flooding regime. In subalpine areas, sphagnum moss and wet heath plants grow in damp low lying areas, whereas across the Murray River floodplain, billabongs and lagoons support species such as cumbungi, common reed and spiny rush. In seasonally flooded low lying black box depressions lignum or nitre goosefoot occurs, and cane grass swamps occur throughout the central and western parts of the catchment.

Matthew Herring

5 Species accounts

This chapter provides accounts of the 80 reptile species known to occur in the NSW Murray catchment. The reptiles featured are grouped by family. At the beginning of the descriptions of each family, we discuss the number of genera and species within that group in Australia, followed by the number in the NSW Murray catchment. We also discuss broad ecological differences between the members within each family. Freshwater turtles in the Family Cheluidae are presented first, followed by the five lizard families and then the three snake families. The lizard and snake families are presented in the following order: (1) Gekkonidae - geckos, (2) Pygopodidae - legless lizards, (3) Varanidae - goannas/monitors, (4) Agamidae - dragons, (5) Scincidae - skinks, (6) Typhlopidae - blind snakes, (7) Pythonidae - pythons, and (8) Elapidae - front-fanged venomous snakes. Within each family, reptile species are presented in alphabetical order according to their Latin name. Similar looking and difficult to identify skink species are discussed in Chapter 6.

Definitions

Common name: Common names listed in this book are based on Swan *et al.* (2004) and Wilson and Swan (2008). The preferred common name is listed first.

Scientific name: Nomenclature in this book is based on Wilson and Swan (2008), with the exception of *Cryptoblepharus pannosus* (= *carnabyi*) (after Horner 2007), *Lerista timida* (= *muelleri*) (after Hutchinson 2008) and *Liopholis* (= *Egernia*) (after Gardner *et al.* 2008). The scientific names of animals are a two-part (binomial) classification system. The first is the generic name (genus) comprising a group of closely related animals; the second is the specific name (the name of the actual species) and it distinguishes it from any other animal in the world. Subspecies (ssp.) names are included where relevant.

Identifying features: We first describe body colour and then dominant markings followed by unique features that readily distinguish species from each other. We have purposely refrained from using technical terms and specific features of scale arrangement. However, in describing some skink species, the use of technical terms relating to the position of particular body markings was unavoidable. The following terms are used: (1) vertebral = referring to the middle of the back; (2) lateral = referring to the flank; and (3) dorsolateral = referring to the junction of the flank and back.

Snout–vent length: A key distinguishing feature of some reptile species is snout-vent length. This measurement is taken from the tip of the snout to the base of the tail or vent and is based on the maximum length of adults provided in Swan *et al.* (2004).

Similar species: Some species look alike. When more than one similar looking species is found broadly in the same environment we list the key features that will help distinguish between them.

Life-form: Reptiles have different body shapes depending on the environment in which they live. For example, species that burrowing through loose soil have elongated bodies and reduced limbs and are termed 'fossorial'. Species that possess well-developed limbs and are capable of rapid movements mostly live on the ground and are termed 'terrestrial'. Species that live on rock outcrops or on trees and shelter in narrow crevices have flattened bodies, elongated limbs and long claws. These species are termed 'arboreal' if they live on trees and 'saxicolous' if they live on rock outcrops. Yet other reptile species are closely associated with water bodies and are termed 'aquatic' or 'semi-aquatic'.

Reproductive mode: There are two modes of reproduction in Australia reptiles. Some lay eggs (oviparous) and some give birth to live young (viviparous). Generally, species that occur in arid environments lay eggs whereas many of those reptile species that occur in colder climates give birth to live young. One way that experienced herpetologists distinguish between viviparous and oviparous snakes is to inspect the scale arrangement of the underside of the tail (subcaudal scales). Those species that have paired subcaudal scales are oviparous, whereas those that have single subcaudal scales are viviparous.

Activity: Most species are either active during the day (diurnal) or by night (nocturnal). Nocturnal species generally need to find shelter-sites that not only provide protection from predators but are also thermally suitable to enable them to attain preferred body temperatures, allowing them to forage well into the night. Fossorial (soil-dwelling) species elevate their body temperature through contact with warm substrates, terrestrial (ground-dwelling) species regulate their body temperature by shuttling between sun and shade and saxicolous (rock-dwelling) species spend periods basking in direct sunlight.

Distribution: We define the distribution for each species based on the number of vegetation types they readily occupy within the Murray catchment. Species that are termed 'restricted' occur in one vegetation type. 'Localised' species are restricted to two or three vegetation types in either the eastern or western part of the catchment, and 'widespread' species are known to occur in a broad range of vegetation types across the catchment.

Shelter-sites: Shelter-sites are areas that reptile species retreat to when they feel threatened or when temperatures are unsuitable for maintaining activity. The most commonly used shelter-sites are listed for each species. Many species share similar shelter-sites such as fallen timber and surface rocks, and many species will occupy the same shelter-site on a repeated long-term basis. Understanding what constitutes a shelter-site for different species is an important part in taking measures to conserve them.

Vegetation type: We list the broad vegetation types (described in Chapter 4) each species has been recorded in within the Murray catchment. The vegetation types in bold text refer to those most commonly associated with each species.

Conservation status: We define the conservation status for each species based on the number of location records and their relative abundance. Rare species are known from less than 10 locations, uncommon species are known from between 10 and 100 locations and common species are known from more than 100 locations. However, assigning a conservation status to many species is problematic as: 1) there are very few formal records for many species in the wildlife atlas databases; 2) there has been only very limited research on reptiles in the Murray catchment. Therefore, baseline data on distribution patterns and relative abundance of the majority of reptile species are often lacking; 3) many species are extremely cryptic and are only active during certain weather conditions making their detection difficult; and 4) a range of survey techniques need to be used over long periods of time to detect many species in a given area. Thus, the conservation status we provide for some species may change with future surveys.

Management: We provide a list of environments, habitat resources or threatening processes that are important to each of the 80 species featured in this book. Management actions that relate to each theme are described in detail in Chapter 3 and if adopted would be beneficial in the conservation of reptiles in production landscapes. Those species sharing a similar habitat will benefit from similar management actions. However, not all species will respond equally to the same actions. This is because some species, for example, have better dispersal abilities than others and can move readily between restored areas.

Side-necked freshwater turtles

Family Cheluidae

The names 'tortoise' and 'turtle' are often used synonymously, although the term 'freshwater turtle' is preferred, thereby distinguishing them from marine species and the completely land-based tortoises of Europe, Africa and the Americas. Australia has 26 species of freshwater turtle which belong in two families, Cheluidae and Carettochelydidae (which contains a single species – the pig-nosed turtle of the Northern Territory). There are seven genera in the family Cheluidae of which two genera *Chelodina* and *Emydura* account for half of all known species in Australia. Three species occur in the Murray catchment and co-occur within the major rivers including the Murray, Wakool and Edward river systems.

The broad-shelled turtle prefers deep, turbid water and is a specialised carnivore, preying on crayfish, prawns and small fish. It is the least known and recorded of the three species in the region. The long-necked turtle is an opportunistic omnivore and feeds on fish carrion and aquatic insects. It is the most widespread and commonly encountered species in the region and can be extremely abundant in farm dams and irrigation channels. The Murray turtle is also a generalist omnivore, although much of its diet includes a high proportion of rotting vegetation and algae. It is often encountered basking on emergent logs in large lakes and sewerage treatment ponds.

Cheluidae

Broad-shelled turtle

Chelodina expansa

Nick Edards

Identifying features: This is an extremely large turtle with an elongated neck and a broad, flattened shell. It has no facial markings or black-edged markings on the belly scales (scutes).

Carapace length: 480 mm.

Similar species: The much smaller long-necked turtle has black-edged markings on the scutes.

Life-form: Aquatic and terrestrial.

Reproductive mode: Oviparous.

Activity: Diurnal.

Distribution: Restricted to major river systems, oxbow lakes and occasionally found in irrigation channels in the west of the Murray catchment.

Shelter-sites: Submerged tree roots and aquatic vegetation.

Vegetation type: River red gum forest and wetlands.

Conservation status: Rare, although seldom encountered, it may be locally abundant in some areas.

Additional notes: The broad-shelled turtle is rarely encountered as is spends most of the time submerged in deep, turbid water. It preys on fish, crayfish and other crustaceans and strikes at its prey in a snake-like action. Little is known about the status of the broad-shelled turtle in the region, but it is thought to be potentially threatened.

Facts: It is usually only females of the broad-shelled turtle that are encountered as they will often travel some distance from a river to deposit their eggs in a self-excavated nesting chamber. It is during this time that adult turtles are at risk of being predated upon by the red fox, or are killed by collisions with cars. In recent years, a number of road kill specimens carrying full-term eggs have been recorded following summer storms in the Albury region. The broad-shelled turtle has the ability to

extract oxygen from water that is drawn into the vent. This enables it to remain submerged for extended periods of time.

Management: Wetlands and exotic animals.

Long-necked turtle, snake-necked turtle

Chelodina longicollis

Hugh McGregor

Identifying features: This is a medium-sized turtle with an elongated neck and black-edged margins on the underbelly scales (scutes).

Carapace length: 250 mm.

Similar species: The much larger broad-shelled turtle lacks black-edged markings on the scutes.

Life-form: Aquatic and terrestrial.

Reproductive mode: Oviparous.

Activity: Diurnal.

Distribution: Widespread in rivers, creeks, wetlands, oxbow lakes, sewerage ponds, irrigation channels and farm dams. This species penetrates further east than any other turtle in the region.

Shelter-sites: Aquatic vegetation.

Vegetation type: River red gum forest, black box woodland and wetlands.

Conservation status: Common. Most populations are considered secure.

Additional notes: The long-necked turtle is the most commonly encountered turtle in the region and is quite at home in small farm dams. During summer rains, it can be encountered crossing roads or in paddocks some distance from water bodies. Like other turtle species that nest on river banks, the eggs are susceptible to being preyed upon by the red fox.

Facts: When water bodies begin to dry up, the long-necked turtle has an innate ability to locate other wetlands, often many

kilometres away. Exactly how these turtles know whether these wetlands contain water or not, no one really knows, but they may be able to detect ultraviolet radiation that is reflected of the surface of the water. This may explain why the long-necked turtle will stand motionless for some time with their neck outstretched and head pointing towards the sky. If the long-necked turtle is picked up, it will often emit a foul smell from glands near the base of the tail.

Management: Wetlands and exotic animals.

Cheluidae

Murray turtle, Macquarie turtle

Emydura macquarii ssp. *macquarii*

Peter Robertson

Identifying features: The neck is much shorter than the length of the shell and it often has a yellow stripe extending from the eye to the check. It lacks black-edged marking on the scutes (belly scales).

Carapace length: 340 mm.

Similar species: None.

Life-form: Aquatic and terrestrial.

Reproductive mode: Oviparous.

Activity: Diurnal.

Distribution: Restricted to major river systems, large lakes and sewage ponds.

Shelter-site: Aquatic vegetation.

Vegetation type: River red gum forest and wetlands.

Conservation status: Uncommon, although some populations are abundant in localised areas.

Additional notes: This species can be extremely abundant in the Murray River system, but like other turtle species, nests are heavily predated upon by the red fox. The Murray turtle deposits its eggs on the bank close to a river making them less susceptible to collisions with motor vehicle. The distinctive eye stripe on many individuals gives them the appearance of smiling, but individuals in some populations may lack this identifying feature.

Facts: The Murray turtle is the most herbivorous of the three turtles in the Murray catchment and consumes large amounts of vegetable material. This may be one reason

why it is extremely abundant in sewage treatment ponds, a place where algae thrives. During extreme periods of drought, many turtle species in the Murray catchment will bury themselves beneath the ground. They remain in a torpor-like state called 'aestivation' until heavy rain occurs. Turtles are long-lived reptiles and can attain ages of 50 years or more.

Management: Wetlands and exotic animals.

Geckos

Family Gekkonidae

Geckos are a cosmopolitan group of lizards that are found in tropical and temperate environments around the world. They can be identified by their large eyes, soft bodies and velvety skin, and are well known for colonising remote islands by drifting on floating mats of vegetation that has been blown out to sea during cyclones and severe storms. Australia has 115 species of gecko which belong to two subfamilies: Gekkoninae and Diplodactylinae. These two groups differ in their origins and condition of their eggs. Species in the Gekkoninae occur widely in other parts of the world and have hard-shelled eggs. In contrast, geckos in the subfamily Diplodactylinae are endemic to Australia, New Zealand and New Caledonia and have soft, parchment-like eggs. Nine species of gecko occur in the Murray catchment. Only two, the southern marbled gecko and the similar looking tree dtella, belong to Gekkoninae. The remaining seven are in the Diplodactylinae.

The majority of geckos in the NSW Murray catchment are found in the western part of the region. Only two species, the southern marbled gecko and stone gecko, survive in cool-temperature environments. The former occurs as far south as Melbourne and is one of the world's most southerly occurring gecko species.

The nine gecko species in the Murray catchment have varied habitat and dietary preferences. As such, only two species (up to three species in mallee and sandhill woodland) are likely to be encountered in any given area. When gecko species occur together, one will often be arboreal and the others terrestrial. This is a type of 'niche' separation and minimises the chance of different gecko species competing directly for the same resources.

Gekkonidae

Southern marbled gecko

Christinus marmoratus

Damian Michael

Identifying features: Body colour light to dark grey with a dark reticulated pattern. All digits are clawed and juveniles often have a series of yellow or orange spots on the tail.

Snout–vent length: 70 mm.

Similar species: The tree dtella lacks a claw on the inside digits of the hands and feet.

Life-form: Arboreal, saxicolous and terrestrial.

Reproductive mode: Oviparous.

Activity: Nocturnal.

Distribution: Widespread across the region except for alpine areas.

Shelter-sites: Exfoliating bark, fallen timber, rock crevices and human dwellings.

Vegetation type: Dry forest, **grassy box woodland**, sandhill woodland, black box woodland, **river red gum forest** and mallee woodland.

Conservation status: Common.

Additional notes: The southern marbled gecko is without a doubt the most abundant and commonly encountered gecko in the region. It is usually the species observed climbing across walls and windows on warm nights to catch insects near light sources. Like all geckos, the southern marbled gecko usually produces two eggs per clutch, which are laid beneath fallen timber and debris. Juvenile geckos hatch during early to mid-summer and disperse to neighbouring trees. They attain sexually maturity at around two years of age.

Facts: Mature females often have enlarged, white bulges on each side of the neck called endolymphatic sacs. These modified glands, which are clearly visible through the translucent skin, often give the southern marbled gecko a 'hammer-head'-like appearance. These glands are thought to aid in the production of calcium, used in the development of eggs.

Management: Rocky outcrops, tree plantings, fallen timber, surface rocks and large mature trees.

Damian Michael

Gekkonidae

Tessellated gecko

Diplodactylus tessellatus

Damian Michael

Identifying features: Body colour grey to reddish-brown with blotches and paired spots along the back. The tail is short and plump.

Snout–vent length: 50 mm.

Similar species: None.

Life-form: Terrestrial.

Reproductive mode: Oviparous.

Activity: Nocturnal.

Distribution: Widespread in the western part of the catchment west of Finley.

Shelter-sites: Soil cracks, fallen timber and ground debris.

Vegetation type: Black box woodland, **boree woodland** and **chenopod shrubland**.

Conservation status: Uncommon, although potentially abundant in some areas.

Additional notes: The tessellated gecko usually shelters during the day within soil cracks, although occasionally individuals live beneath leaf litter. They can often be found when spotlighting on warm nights.

Facts: Studying reptiles can often be difficult as more than one survey method is needed to effectively record a range of species in an area. A study of native grasslands in northern Victoria found that using old fence posts was an effective way of sampling and monitoring grassland reptiles. During one year, 12 species were recorded beneath the fence posts including the tessellated gecko. Old fence posts can be strategically placed on the ground in small

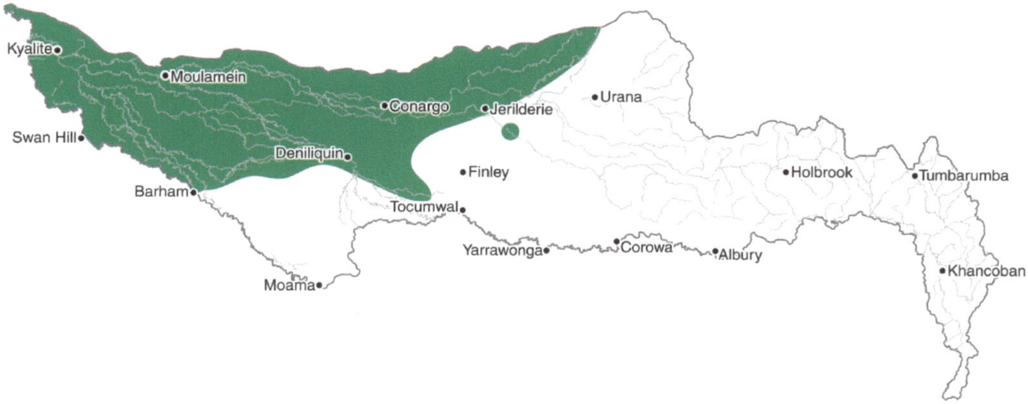

clusters along re-established fence lines. This is an effective way of recycling material, monitoring wildlife and creating habitat for small animals in farming areas.

Management: Soil structure and integrity, leaf litter and woody debris, and shrubs.

Damian Michael

Gekkonidae

Eastern stone gecko

Diplodactylus vittatus

Identifying features: Body colour reddish-brown with a zigzag or 'zipper'-like pattern along the body and onto the tail. The tail is short and plump.

Snout–vent length: 50 mm.

Life-form: Terrestrial.

Reproductive mode: Oviparous.

Activity: Nocturnal.

Distribution: Widespread but patchily distributed across the region.

Shelter-sites: Surface rocks, exfoliating bark of fallen dead trees, fallen timber and ground debris.

Vegetation type: **Dry forest**, grassy box woodland, sandhill woodland and boree woodland.

Conservation status: Rare in the west and uncommon in the east.

Additional notes: The eastern stone gecko is often found in pairs or small aggregations. This may be a result of limited shelter-sites or it could indicate breeding behaviour. Due to a tolerance for low temperature, it penetrates further east than any other gecko in the region. Considerable variation in the body pattern exists among the eastern stone gecko and it is possible that this species is a complex of several undescribed subspecies.

Facts: In the east of the region, the eastern stone gecko often shelters beneath a favourite set of rocks. Well-used shelter-sites will contain a number of shed skins suggesting long-term residency. These rocks are usually relatively thin and are

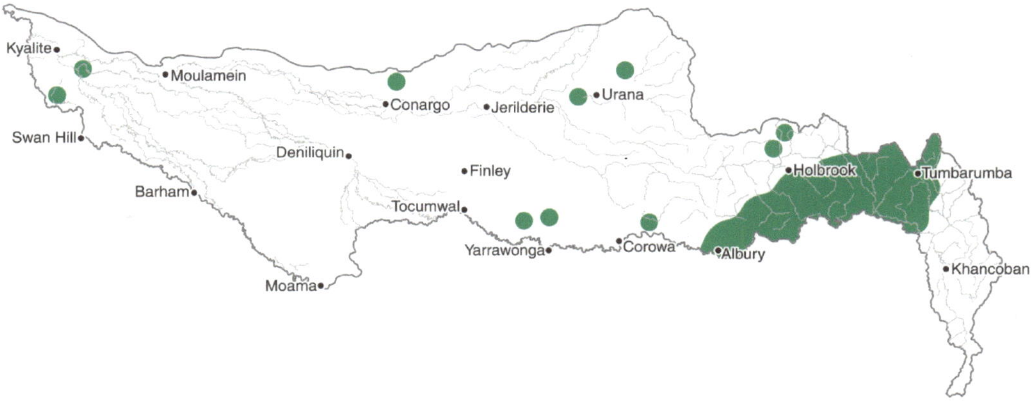

exposed to direct sunlight which allows the eastern stone gecko to reach a preferred body temperature beneath the protection of cover. The variation in thermal conditions provided by rocks that lie on top of one another enables the eastern stone gecko to regulate its body temperature and forage well into the night during relatively cool conditions. Illegal bush rock removal can have a significant impact on rock-dwelling species because once this type of habitat is destroyed it is impossible to replace.

Management: Rocky outcrops, surface rocks, and fallen timber and dead trees.

Damian Michael

Gekkonidae

Tree dtella

Gehyra variegata

Lachlan McBurney

Identifying features: Body colour dark grey with dark checked body pattern. Tail is thin and tapering and inner claw on 'thumb' digit lacking on hands and feet.

Snout–vent length: 54 mm.

Similar species: The southern marbled gecko has claws on all digits.

Life-form: Arboreal and saxicolous.

Reproductive mode: Oviparous.

Activity: Nocturnal.

Distribution: Restricted. First recorded in 2003 in the Murray catchment and currently known only from a travelling stock reserve north of Kyalite in the west, Moulamein and Galore Hill in the north of the catchment.

Shelter-sites: Exfoliating bark and surface rocks.

Vegetation type: Grassy box woodland and mallee woodland.

Conservation status: Rare.

Additional notes: The tree dtella is one of Australia's most commonly encountered reptiles across its geographical range, which spans almost the entire continent.

Facts: The tree dtella is able to persist in relatively small patches of habitat in agricultural areas. This is because the tree dtella is able to alter its habitat use in response to changes in the environment. Populations in nature reserves are more commonly found sheltering in shrubs, whereas in populations surviving in small fragmented remnant vegetation use logs and eucalypt trees more frequently. This behavioural trait is an important characteristic that enables many species to persist in landscapes where grazing pressure can change vegetation structure and the type of plants that occur. The tree dtella may be rare in the region because it has to compete for resources with the more common southern marbled gecko.

Management: Fallen timber and dead tree, shrubs and large mature trees.

Peter Robertson

Gekkonidae

Gibber gecko
Lucasium byrnei

David Webb

Identifying features: Body colour light brown to reddish-brown with a series of dark 'hourglass'-shaped markings on body and tail. The tail is long and slender.

Snout–vent length: 55 mm.

Similar species: None.

Life-form: Terrestrial.

Reproductive mode: Oviparous.

Activity: Nocturnal.

Distribution: Restricted. First recorded in 2005 in the Murray catchment and known only from a single population located between Moulamein and Kyalite.

Shelter-sites: Soil crevices, spider holes and leaf litter.

Vegetation type: Chenopod shrubland.

Conservation status: Rare.

Additional notes: During the day, the gibber gecko shelters beneath leaf litter, in burrows and ground cracks emerging at night to actively forage for insects. It lays two eggs per clutch.

Facts: The gibber gecko gets its name from the stony 'gibber' plains of inland Australia. Gibber is an Aboriginal word meaning stone. The gibber gecko is not solely restricted to the gibber plains, and is just as likely to be found in mallee woodland, chenopod shrubland and sandy deserts. Like other geckos, the gibber gecko does not have eyelids. Instead, they have a single clear spectacle which acts as a protective scale covering its eyes. When vision becomes obscured by dust, it simply uses its fleshy tongue to wipe the eyes clean.

Management: Soil structure and integrity, leaf litter and fine woody debris, and surface rocks.

Kyalite
Moulamein
Conargo
Jerilderie
Urana
Swan Hill
Deniliquin
Finley
Holbrook
Tumbarumba
Barham
Tocumwal
Yarrawonga
Corowa
Albury
Khancoban
Moama

David Webb

Gekkonidae

Beaded gecko

Lucasium damaeum

David Webb

Identifying features: Body colour red to brown with a pale vertebral stripe and small spots along the body. The tail is thin and tapering and eye lid white.

Snout–vent length: 55 mm.

Similar species: The beaked gecko has a more pronounced wedge-shaped snout and lacks a vertebral stripe.

Life-form: Terrestrial.

Reproductive mode: Oviparous.

Activity: Nocturnal.

Distribution: Restricted. First recorded in 2003 in the Murray catchment and known only from a single population near Koraleigh.

Shelter-sites: Spider holes and self-made burrows.

Vegetation type: Mallee woodland.

Conservation status: Rare.

Additional notes: The beaded gecko is a common species in Australia. It is closely associated with spinifex grass where it constructs shallow burrows at the base of the plant. At night, the beaded gecko forages in open areas in search of insects and spiders. It lays two eggs per clutch.

Facts: Australian geckos are divided into two groups: the subfamilies Gekkoninae and Diplodactylinae. The beaded gecko belongs to the latter group which consists of approximately 84 species, all of which are thought to have evolved in Australia. They are the largest subfamily comprising 75% of all gecko species in Australia. They are thought to be closely related to the legless or flap-footed lizards from the Family Pygopodidae, despite having completely different body forms (see section on legless lizards).

Management: Native grass (including spinifex), leaf litter and fine woody debris.

Kyalite
Moulamein
Conargo
Jerilderie
Urana
Swan Hill
Deniliquin
Finley
Holbrook
Tumbarumba
Barham
Tocumwal
Yarrawonga
Corowa
Albury
Khancoban
Moama

Gekkonidae

Beaked gecko

Rhynchoedura ornata

David Webb

Identifying features: Body colour reddish-brown with a series of pale spots along the body and distinctive beaked snout.

Snout–vent length: 54 mm.

Similar species: The beaded gecko has a less pronounced wedge-shaped snout and a vertebral stripe.

Life-form: Terrestrial.

Reproductive mode: Oviparous.

Activity: Nocturnal.

Distribution: Restricted. First recorded in 2005 in the Murray catchment and known only from a single population between Moulamein and Kyalite.

Shelter-sites: Spider holes, soil cracks and leaf litter.

Vegetation type: Mallee woodland and sandhill woodland.

Conservation status: Rare.

Additional notes: The beaked gecko has a specialised diet and is one of the few species that feeds principally on termites. Such a high degree of dietary specialisation is possible only because of the colonial nature of termites. The costs of locating termites are outweighed by the benefits of massive termite populations throughout Australia.

Facts: The beaked gecko is one of a suite of species that utilises vertical spider holes to shelter in during the day. Being arid-adapted, the species is tolerant of high ambient temperatures and is capable of thermoregulating within invertebrate tunnels. It moves up and down these tunnels depending on how hot it is outside. During days of extreme heat, it will venture deep below ground, whereas on cooler days it will access heat closer to the surface.

Management: Soil structure and integrity, leaf litter and fine woody debris.

Kyalite
Moulamein
Conargo
Jerilderie
Urana
Swan Hill
Deniliquin
Finley
Holbrook
Tumbarumba
Barham
Tocumwal
Yarrawonga
Corowa
Albury
Khancoban
Moama

David Webb

Gekkonidae

Eastern spiny-tailed gecko

Strophurus intermedius

Damian Michael

Identifying features: Body colour pale to dark grey with dark-flecked irregular markings. Two (occasionally four) rows of conical spines along body and tail.

Snout–vent length: 60 mm.

Similar species: None.

Life-form: Arboreal.

Reproductive mode: Oviparous.

Activity: Nocturnal but occasionally found basking during the day on the branches of shrubs.

Distribution: Widespread but patchily distributed. Predominantly found throughout the central and northern parts of the region, recorded as far east as Morgan's Ridge near Holbrook.

Shelter-sites: Exfoliating bark, dead tree stumps, fallen timber and branches of shrubs.

Vegetation type: Grassy box woodland, **sandhill woodland** and mallee woodland.

Conservation status: Uncommon, although some populations are abundant and considered secure.

Additional notes: The eastern spiny-tailed gecko has a close association with cypress pine in the Murray catchment. It can be commonly found sheltering behind exfoliating bark or in the crevices of old dead trees and tree stumps. The relative abundance of eastern spiny-tailed geckos in this plant community may be a reflection of past management practices. Areas that were heavily logged and have since regenerated contain fewer individuals than stands of multi-aged trees, tree stumps and dead trees.

Facts: When the eastern spiny-tailed gecko feels threatened, it will raise its tail in defence displaying a row of spines. As a last

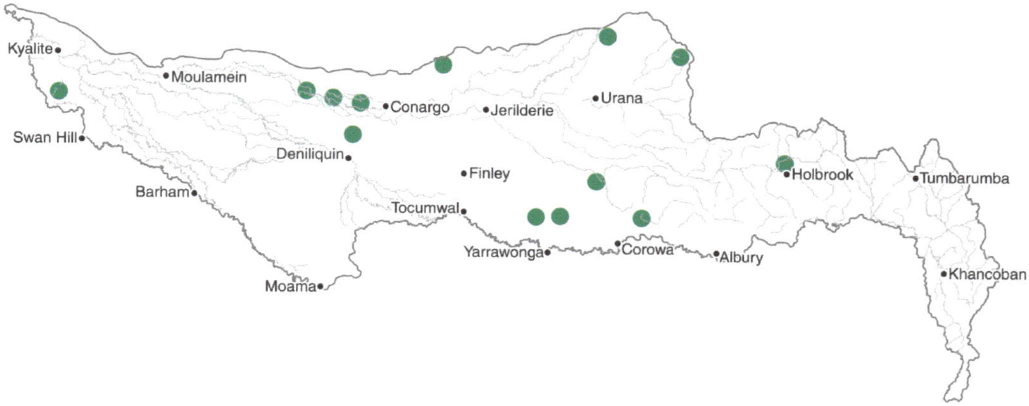

resort, it will exude a sticky fluid from glands located beneath these spines. Many predators are deterred by the taste of this substance and will often reject the gecko in search of a more appealing meal.

Management: Fallen timber and dead trees, and shrubs.

Damian Michael

Gekkonidae

Thick-tailed gecko, barking gecko

Underwoodisaurus milii

Lachlan McBurney

Identifying features: Body colour red to brown with bands of white spots along body and tail. Original tail is leaf-shaped and patterned. Regenerated tails lack markings.

Snout–vent length: 65 mm.

Similar species: None.

Life-form: Terrestrial.

Reproductive mode: Oviparous.

Activity: Nocturnal.

Distribution: Restricted. First recorded in 2008 in the Murray catchment and known only from a single record near Kyalite.

Shelter-sites: Burrows and ground debris.

Vegetation type: Mallee woodland.

Conservation status: Rare.

Additional notes: The thick-tailed gecko is one of the most strikingly coloured of all the geckos in the region. Some people liken its body pattern to that of indigenous artwork because of its bright colours and spots. It is one the largest terrestrial geckos in Australia and stalks its prey at night. It will even eat small reptiles. It is often encountered in pairs or small groups beneath surface rocks. In some parts of its range, it will shelter in disused rabbit burrows.

Facts: When threatened, the thick-tailed gecko will stand tall on its skinny legs, raise its tail above its back and make small lunging advances towards the attacker. It will also emit a croaky barking sound which gives it the other common name of barking gecko.

Management: Soil structure and integrity, leaf litter and fine woody debris, and surface rocks.

Damian Michael

Legless lizards

Family Pygopodidae

The term 'legless lizard' is not technically accurate as the name pygopod means 'flap-footed', which refers to the vestigial hind limbs present in many species. This feature, along with obvious ear-openings (present in most species), a fleshy tongue and an ability to vocalise help scientists distinguish them from snakes. They also have much longer tails than snakes, often accounting for more than three-quarters of the total body length. Legless lizards are almost uniquely Australian, although one species extends to Papua New Guinea and another species is endemic to that country. Australia has 41 species of legless lizard which belong in seven genera. Three genera, *Aprasia, Delma* and *Pygopus* account for over 90% of all species. The remaining four genera contain a single species. The four legless lizard species that occur in the Murray catchment each belong to a different genus.

The four legless lizards in the Murray catchment have broadly different habitat preferences. Therefore, it is unlikely that more than two species will be encountered in the same area. The only exception to this is in the Albury region, where three species (pink-tailed worm-lizard, olive legless lizard and Burton's snake-lizard) have been recorded from the Nail Can Hill Flora and Fauna Reserve. Two of the species in the catchment (pink-tailed worm-lizard and Burton's snake-lizard) are associated with rocky environments, while the other two species (olive legless lizard and eastern hooded scaly-foot) are associated with grassy woodlands or derived grasslands.

Pink-tailed worm-lizard

Aprasia parapulchella

Identifying features: Body colour light grey with rows of fine, black lines along body. Head dark, tail tip pinkish and no external ear-openings.

Snout–vent length: 140 mm.

Similar species: The olive legless lizard has obvious external ear-openings and hind-limb flaps.

Life-form: Fossorial.

Reproductive mode: Oviparous.

Activity: Nocturnal.

Distribution: Restricted. Known from four locations in the Murray catchment. A population on Nail Can Hill in Albury, a single record from both Goombargana Hill and Howlong Hills, and several individuals from between Yarra Creek and Holbrook. In Victoria, it is likely to occur in similar habitat around Wodonga, but is currently only known from the Greater Bendigo region.

Shelter-sites: Metamorphic and granite rocks and within the nests of *Iridomyrmex* ant species.

Vegetation type: Grassy box woodland.

Conservation status: Uncommon, most populations are considered secure, although the pink-tailed worm-lizard is listed as 'vulnerable to extinction' in NSW.

Additional notes: The pink-tailed worm-lizard is a secretive lizard that is usually detected beneath shallow surface rocks between spring and autumn. The spring period is when the ant species it shares its home with are brooding eggs in small

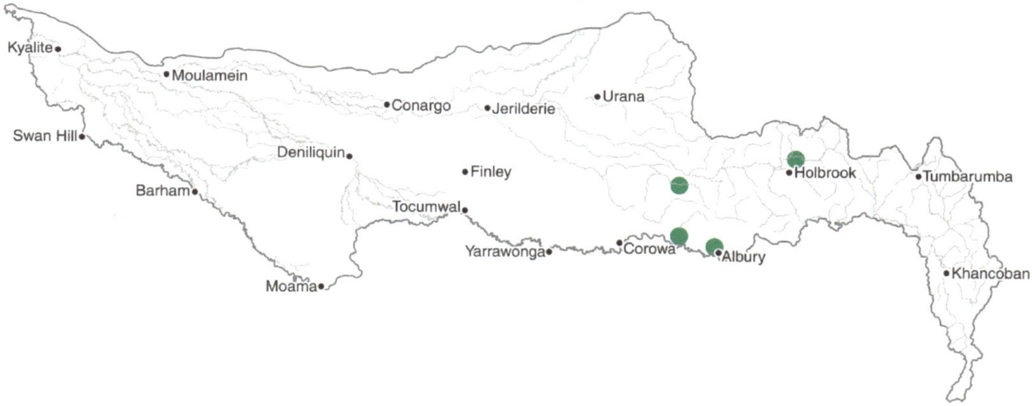

chambers. Small aggregations have been observed beneath rocks and the species is also known to shelter beneath discarded corrugated iron. These lizards are active during warm nights and have been found up to 50 m away from surface rock habitat.

Facts: If handled an individual will coil its body into a knot and attempt to twist free, a behaviour that may help them escape from predators.

Management: Soil structure and integrity, native grass, rocky outcrops and surface rocks.

Damian Michael

Pygopodidae

Olive legless lizard

Delma inornata

Damian Michael

Identifying features: Body colour light grey to olive-green with a yellow tinge to throat and obvious external ear-openings.

Snout–vent length: 133 mm.

Similar species: The pink-tailed worm-lizard and small snakes lack external ear-openings and hind-limb flaps.

Life-form: Terrestrial.

Reproductive mode: Oviparous.

Activity: Nocturnal, but occasionally seen basking during the day on top of grass tussocks.

Distribution: Widespread but absent from the far eastern part of the catchment.

Shelter-sites: Grass tussocks, fallen timber, fallen shrubs, litter mats and surface rocks.

Vegetation type: Grassy box woodland, sandhill woodland, black box woodland, chenopod shrubland, **boree woodland** and mallee woodland.

Conservation status: Common.

Additional notes: The olive legless lizard is often found in open grassy areas, although it lays two eggs beneath fallen timber and surface rocks. It is often found coiled beneath embedded fallen timber and when disturbed, escapes by springing into the air before quickly disappearing into the grass.

Facts: Recent surveys in the Murray catchment found the olive legless lizard to be common in tree plantings, especially if these areas are ungrazed and shrubs are left to die naturally. A long-term study in the SWS bioregion involving the use of corrugated iron, terracotta roofing tiles and wooden palings to survey reptiles indicates that the olive legless lizard may be increasing in abundance, especially in tree plantings. This suggests that the species has some ability to disperse across grazed landscapes and colonise restored areas.

Management: Native grass, tree plantings, surface rocks, fallen timber and dead tree, and shrubs.

Damian Michael

Burton's snake-lizard

Lialis burtonis

Clare McCutcheon

Identifying features: Body colour light grey with or without fine, black markings. Wedge-shaped snout distinguishes it from snakes and other legless lizards.

Snout–vent length: 290 mm.

Similar species: None.

Life-form: Terrestrial.

Reproductive mode: Oviparous.

Activity: Nocturnal but occasionally encountered during the day in overcast conditions.

Distribution: Restricted. Known only from Nail Can Hill Flora and Fauna Reserve in Albury and near Tumbarumba in the east of the catchment.

Shelter-sites: Fallen timber and surface rocks.

Vegetation type: Dry forest and grassy box woodland.

Conservation status: Rare.

Additional notes: Burton's snake-lizard is a peculiar looking lizard with a stout body, large eyes and a long, wedged-shaped head. It is a specialist carnivore that feeds principally on fast moving lizards. It is considered to be one of Australia's most widespread and common species. In the Murray catchment, Burton's snake-lizard lacks a distinctive body pattern. However, it is possible that striped forms may also occur.

Facts: Burton's snake-lizard is unusual among reptiles in that it has the ability to move its skull bones at a certain point, a process called cranial kinesis. This gives the mouth greater mobility and enables the 'hinged' jaws to wrap around relatively large lizards. Prey items are firmly held by teeth that are slightly curved and face towards the back of the mouth.

Management: Rocky outcrops, surface rocks, and fallen timber and dead trees.

Kyalite
Moulamein
Conargo
Jerilderie
Urana
Swan Hill
Deniliquin
Finley
Holbrook
Tumbarumba
Barham
Tocumwal
Yarrawonga
Corowa
Albury
Khancoban
Moama

David Webb

Pygopodidae

Eastern hooded scaly-foot

Pygopus schraderi

David Webb

Identifying features: Body colour reddish-brown with black facial markings and obvious hind-limb flaps.

Snout–vent length: 198 mm.

Similar species: The curl snake has a plain brown body and juvenile eastern brown snakes have a dark head and neck band. All snakes lack ear-openings and hind-limb flaps.

Life-form: Terrestrial.

Reproductive mode: Oviparous.

Activity: Nocturnal.

Distribution: Localised. Predominantly found between Conargo and Jerilderie in the northern part of the catchment.

Shelter-sites: Spider holes, soil cracks and fallen timber.

Vegetation type: Boree woodland (often with a grassy ground layer).

Conservation status: Rare, although potentially common in areas supporting native grassland, especially the region bordering the Murrumbidgee catchment in the north.

Additional notes: The eastern hooded scaly-foot is most often encountered by spotlighting on warm nights in native grasslands and grassy boree woodlands. During the day, discarded skins may be found protruding from tunnels in the ground made by spiders and king crickets.

Facts: Many harmless species of reptile have head and neck markings which mimic venomous snakes. The eastern hooded scaly-foot is one example, but goes a step further by exhibiting defensive behaviour similar to the threat display of the eastern brown snake. When threatened, it will raise its fore body off the ground, compress its neck, flick its tongue, and even attempt to strike. Snake mimicry is an important

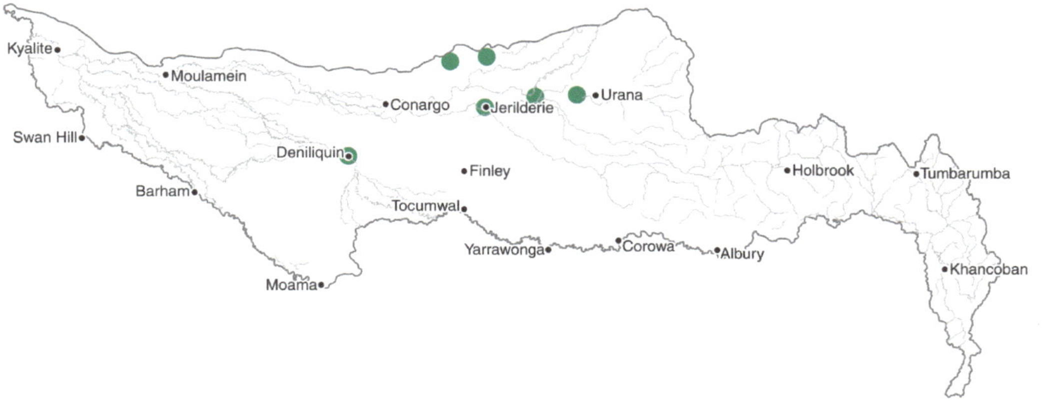

part of its survival strategy. Unfortunately, this behaviour is so impressive that this harmless species is mistaken for a venomous snake and killed. The physical difference between most legless lizards and snakes is the fleshy tongue, external ear openings and hind-limb flaps. Venomous snakes have forked tongues, no external ear openings, and no external limbs.

Management: Soil structure and integrity, native grass, and fallen timber and dead trees.

David Webb

Goannas/Monitors

Family Varanidae

The word goanna is though to be a corruption of the word 'iguana' which is a similar looking reptile from the Americas. Goannas are also called monitor lizards, referring to their habit of standing on their back legs and scanning the surrounding landscape when they perceive a threat. These large reptiles are found in Africa, Asia, Australia and some western Pacific Islands. Australia has 27 species, supporting over half of the world's species, all of which belong to the genus *Varanus*. Goannas live in all states of Australia except Tasmania and can be found in a variety of environments including rainforests, deserts, open woodlands and wetlands. Goannas in Australia can be broadly grouped according to the habitat in which they live. These included saxicolous species (rock-dwelling), arboreal species (tree-dwelling), terrestrial species (ground-dwelling) and semi-aquatic species.

Three species of goanna are found in the Murray catchment. The sand goanna and lace monitor are both widespread species that can be found in similar habitats across most of the Murray catchment. When disturbed, the lace monitor will often climb the nearest tree, whereas the sand goanna usually escapes by retreating into a burrow. However, both species are capable of burrowing and climbing. The heath monitor has a wide distribution in southern Australia, although it only occurs in the far eastern part of the Murray catchment. It is the most cold adapted of all the three species in the catchment and is also the least known.

Varanidae

Sand goanna, sand monitor, Gould's monitor

Varanus gouldii ssp. *gouldii*

David Webb

Identifying features: Body colour pale yellow to dark reddish-brown with light and dark spots forming transverse bands along body. Pale-edged black stripe behind eye. Tail has combination of narrow bands which terminate in a single broad yellow band.

Total length: 1.6 m.

Similar species: The lace monitor has one or more broad yellow bands along the tail and the heath monitor has multiple narrow bands along the tail.

Life-form: Terrestrial but capable of climbing trees.

Reproductive mode: Oviparous.

Activity: Diurnal.

Distribution: Widespread. Not found east of the Hume Highway.

Shelter-sites: Hollow logs, rabbit burrows and self-made burrows.

Vegetation type: Sandhill woodland and mallee woodland.

Conservation status: Common and secure, although populations have declined in the central and eastern part of the Murray catchment.

Additional notes: The sand goanna is sometimes referred to as the racehorse goanna because of its agility and exceptional speed when chasing prey or fleeing from danger. When threatened, it will usually disappear into the nearest burrow, staying hidden for some period of time before reappearing. Often, a second concealed exit will be used as an alternative escape route.

Facts: Anecdotal evidence suggests the sand goanna has declined in some parts of the SWS bioregion, although they may have benefited from the arrival of the introduced rabbit in other parts of the Murray catchment. This is because the species utilises

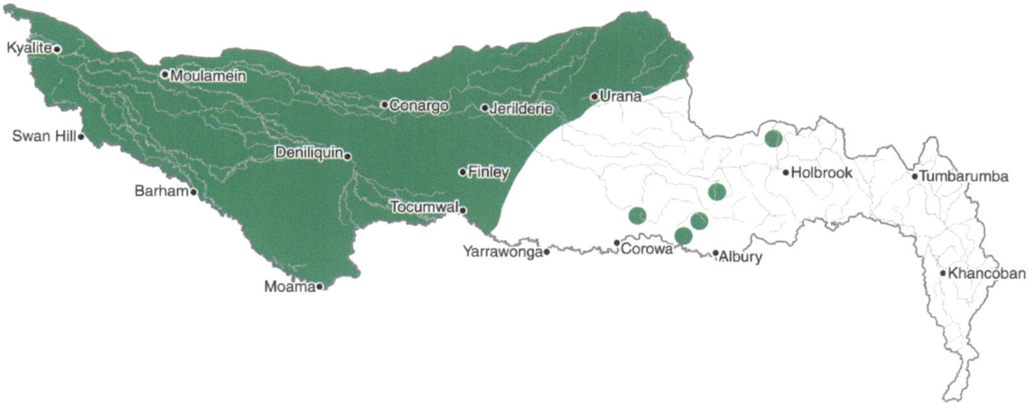

rabbit burrows for shelter and also feeds on rabbits. Rabbits constitute a large proportion of their diet. Because of their fondness for dead animals, goannas are often attracted to road kill, and despite their ability to move quickly, are often killed themselves.

Management: Fallen timber and dead trees, and exotic animals.

Peter Robertson

Heath monitor, Rosenberg's goanna

Varanus rosenbergi

Damian Michael

Identifying features: Body colour dark grey with narrow, pale cross bands along the body and tail. Juveniles have a brightly coloured orange flush to the body.

Total length: 1.3 m.

Similar species: The lace monitor has one or more broad yellow bands along the tail. The sand monitor has a combination of narrow bands which terminate in a single, broad yellow band on the tail and a pale-edged, black stripe behind the eye.

Life-form: Terrestrial.

Reproductive mode: Oviparous.

Activity: Diurnal.

Distribution: Restricted. Known only from Kosciuszko National Park in the vicinity of Khancoban.

Shelter-sites: Hollow trees and logs.

Vegetation type: Montane forest and dry forest.

Conservation status: Rare. The species is listed as 'vulnerable to extinction' in NSW.

Additional notes: The heath monitor has one of the most fragmented distributions of any monitor lizard in Australia. There are isolated populations in south-western Western Australia, on the Eyre Peninsula in South Australia, Kangaroo Island, western Victoria, southern and coastal New South Wales, and the Australian Capital Territory. These populations were once connected during a time when Australia was cooler and tree cover was more continuous. As the continent became more arid, populations of the cold-adapted heath monitor became isolated from each other.

Facts: The heath monitor is well adapted to cold environments and often lays eggs in termite mounds. After a female excavates a hole and deposits her eggs, termites seal the opening creating a thermally stable incubation chamber. When the young hatch, they dig themselves out of the nest chamber,

emerging simultaneously in a similar manner to other reptiles that lay large clutches such as turtles and crocodiles.

Management: Fallen timber and dead trees, large mature trees and exotic animals.

Varanidae

Lace monitor, Tree goanna

Varanus varius

Damian Michael

Identifying features: Two colour morphs of the lace monitor occur in the Murray catchment. One form has narrow body bands and broad tail bands. The other form has broad, black and yellow body bands and is sometimes called Bell's phase. Both colour morphs can be present in the same clutch of eggs.

Total length: 2.1 m.

Similar species: The heath monitor has narrow bands along the tail. The sand monitor has a combination of narrow bands which terminate in a single, broad yellow band on the tail.

Life-form: Arboreal and terrestrial.

Reproductive mode: Oviparous.

Activity: Diurnal.

Distribution: Widespread except for the far eastern part of the catchment.

Shelter-sites: Hollow trees, logs and rabbit burrows.

Vegetation type: Dry forest, **grassy box woodland**, sandhill woodland, black box woodland, **river red gum forest** and mallee woodland.

Conservation status: Common and secure but known to have declined in some parts of the catchment.

Additional notes: The lace monitor is the second largest lizard in Australia and the fourth largest monitor lizard in the world. The top three are the Komodo dragon, from the Komodo Islands, the water monitor from South-East Asia and the Australian perentie.

Facts: Ancestors of the varanid family are thought to have given rise to snakes. Both posses a forked tongue and the ability to detect airborne scent particles by flicking out the tongue and inserting it into pits on the roof of the mouth called the Jacobson organ. Recently, the lace monitor has been found to contain toxin-secreting oral glands in the lower jaw. Venom that was extracted from these glands had a significant affect on laboratory rats. People bitten by monitor lizards have reported extreme pain, breathing difficulties and muscle weakness suggesting they had been envenomated.

Management: Fallen timber and dead trees, large mature trees and exotic animals.

Damian Michael

Dragons

Family Agamidae

The word 'dragon' is derived from Greek words that mean 'large serpent' and the active verb 'to see clearly', in reference to their keen sense of vision. There are over 300 species of dragon lizard in the world and they can be found in Africa, Europe, Asia, Australia and western Pacific Islands. They are well adapted to hot environments and reach their greatest diversity in arid and dry tropical parts of Australia. They can be distinguished from other reptiles by their long tails, well-developed legs, wedge-shaped head and rough body texture. Australia is home to 70 species of dragon which belong in thirteen different genera. Two genera, *Ctenophorus* and *Diporiphora* account for over half of all dragons in Australia, although neither of these two groups is represented in the Murray catchment. Five species occur in the Murray catchment, although only one species, the eastern bearded dragon, is widespread and common.

Three of the five dragon species that occur in the Murray catchment are similar in appearance and may be difficult to distinguish. However, it is rare for more than two species to be encountered in the same area. The exception is in the Albury region, where three species (eastern bearded dragon, jacky lizard and nobbi dragon) co-occur on the Nail Can Hill Flora and Fauna Reserve and on the Yambla Range. The other two species occur in the High Country and are easily identified by their body size and different habitat preferences. Dragons will often remain motionless when threatened and rely on their camouflage for protection. This makes them difficult to detect in the wild, although the eastern bearded dragon is often seen basking by roadsides or on tree stumps and fence posts.

Agamidae

Jacky lizard

Amphibolurus muricatus

Damian Michael

Identifying features: Body colour grey with pale grey or white body stripes and tail bands. Mouth lining yellow and body scales rough.

Snout–vent length: 120 mm.

Similar species: The nobbi dragon has a pinkish mouth lining and more uniform body scales and juvenile eastern bearded dragons have spiny 'beards'.

Life-form: Terrestrial and semi-arboreal.

Reproductive mode: Oviparous.

Activity: Diurnal.

Distribution: Localised. Predominantly found east of the Hume Highway but known also from Nail Can Hill in Albury and the Yambla Range.

Shelter-sites: Fallen timber and surface rocks.

Vegetation type: Montane forest, **dry forest** and grassy box woodland.

Conservation status: Common.

Additional notes: The jacky lizard employs a 'sit and wait' foraging tactic and is commonly seen perching on low branches or tree stumps. It scans the surrounding environment for a suitable meal, which might include grasshoppers or beetles. When a prey item is detected, the jacky lizard will jump off its perch and quickly chase it. Once it has consumed the prey, it returns to the same perch and signals to other lizards of the same species by waving its arm in the air. It also communicates by head bobbing which serves to reduce territorial disputes between neighbours.

Facts: The jacky lizard can change body colour depending on air temperature and stress levels. Within minutes it can change

from being dark grey and completely patternless to light grey with bold black and white stripes and blotches.

Management: Rocky outcrops, surface rocks, fallen timber and dead trees, and exotic animals.

Damian Michael

Agamidae

Nobbi dragon

Amphibolurus nobbi ssp. *nobbi*

Damian Michael

Identifying features: Body colour grey to tan-brown with a series of dark blotches and pale stripes along the body. Mouth lining pinkish and body scales are uniformly arranged. Breeding males have a pinkish flush to the base of the tail.

Snout–vent length: 80 mm.

Similar species: The jacky lizard has a yellow mouth lining and uneven body scales. Juveniles of the eastern bearded dragon have spiny 'beards' and small spines along the flanks.

Life-form: Terrestrial.

Reproductive mode: Oviparous.

Activity: Diurnal but recorded being active on warm nights.

Distribution: Localised. Disjunct populations occur in the Murray catchment. In the east of the catchment on Galore Hill, the Yambla Range and Nail Can Hill Flora and Fauna Reserve in Albury; in the west from near Koraleigh.

Shelter-sites: Fallen timber and surface rocks.

Vegetation type: Grassy box woodland and mallee woodland.

Conservation status: Rare, although known populations are considered secure.

Additional notes: During the spring and autumn, the nobbi dragon will congregate in areas with suitable habitat. During a reptile survey in Albury, 12 nobbi dragons were found perched on fallen timber. A similar aggregation was recorded in the same location in autumn. These aggregations may develop as part of breeding activity and it is likely that in the Murray

catchment the nobbi dragon will breed a second time and deposit eggs over winter.

Facts: The nobbi dragon is unusual among dragons in that it has been reported to travel as far as 15 km from breeding areas and has been recorded as being active at night.

Management: Rocky outcrops, surface rocks, fallen timber and dead trees, and exotic animals.

Damian Michael

Gippsland water dragon

Physignathus lesueurii ssp. howitti

David Hunter

Identifying features: Body colour dark grey to olive-green with a series of light and dark cross bands along the body and tail. Breeding males have an orange and blue throat.

Snout–vent length: 245 mm.

Similar species: None.

Life-form: Terrestrial, arboreal and semi-aquatic.

Reproductive mode: Oviparous.

Activity: Diurnal.

Distribution: Restricted to the upper reaches of the Tooma River in the east of the catchment. However, anecdotal records suggest a population may once have occurred in Barmah State Forest in the early 1980s.

Shelter-sites: Branches, vegetation, rock ledges, hollow logs and burrows.

Vegetation type: Montane forest, dry forest and wetlands.

Conservation status: Rare.

Additional notes: The Gippsland water dragon is a powerful swimmer and a skilled climber. It will often bask on overhanging branches along creeks and rivers. If disturbed, it will drop into the water and remain submerged until it feels safe to reappear.

Facts: The water dragon has an unusual method of determining the sex of its offspring – a reproductive strategy called temperature-dependent sex determination. Temperatures within nests that are above 28°C and below 26°C produce mostly

female offspring. Nest temperatures between 26°C and 28°C produce mostly male offspring. The females' ability to select optimal nesting sites is an important behavioural trait which may help them adapt to the effects of climate change. This is because they may be able to continuously adjust their nesting sites as the climate becomes warmer so that male water dragons still account for some of the offspring.

Management: Wetlands, surface rocks, fallen timber and dead trees, and exotic animals.

Agamidae

Eastern bearded dragon

Pogona barbata

Damian Michael

Identifying features: Body colour grey to tan-brown with a dark eye stripe and series of paired pale blotches along the body. The tail has light coloured cross bands and adults have a distinctive, dark coloured, spiny throat and spines along the flanks.

Snout–vent length: 250 mm.

Similar species: The nobbi dragon has a pinkish mouth lining and more uniform body scales. The jacky lizard has a yellow mouth lining and uneven body scales. Both of these species lack spines along the flank.

Life-form: Terrestrial.

Reproductive mode: Oviparous.

Activity: Diurnal.

Distribution: Widespread except for the far eastern part of the Murray catchment.

Shelter-sites: Fallen timber, fence posts, tree stumps and burrows.

Vegetation type: Dry forest, **grassy box woodland**, black box woodland, **sandhill woodland**, **boree woodland,** chenopod shrubland and **mallee woodland**.

Conservation status: Common, although some populations in the eastern part of Murray catchment have declined.

Additional notes: The eastern bearded dragon has one of the largest clutch sizes of any Australian lizard. It can lay up to 30 eggs per clutch and will often produce two clutches per year. It excavates a shallow burrow in which the eggs are laid and then back fills with dirt.

Facts: The eastern bearded dragon is often erroneously called a 'frill-neck lizard',

although this name is reserved for a related species from northern Australia which has a large (soccer ball-sized) fold of skin surrounding its neck. That species is not known from NSW. Like many dragon species, the eastern bearded dragon has a preferred body temperature averaging 35°C. They maintain this temperature by changing their body position and shifting between substrates depending on the ambient temperature.

Management: Fallen timber and dead trees, and exotic animals.

Damian Michael

Agamidae

Mountain dragon, mountain heath dragon

Rankinia diemensis

Identifying features: Body colour grey to reddish-brown with a two rows of semi-circular, pale blotches along the body. Mouth lining is blue and the base of the tail has distinctive spiny scales.

Snout–vent length: 82 mm.

Similar species: The jacky lizard has a yellow mouth lining and uneven body scales.

Life-form: Terrestrial.

Reproductive mode: Oviparous.

Activity: Diurnal.

Distribution: Restricted to Kosciuszko National Park in the east of the Murray catchment.

Shelter-sites: Fallen timber, surface rocks and burrows.

Vegetation type: Alpine woodland and **montane forest**.

Conservation status: Uncommon.

Additional notes: The mountain dragon is unusual among dragon lizards in that it is extremely tolerant of cold conditions. It has a preferred body temperature of less than 30°C, which is relatively low when compared with other dragon species. Because it lives in a cool climate, it uses burrows for shelter during the night. It has been reported to live for as long as ten years in captivity.

Facts: Most dragon species live in hot, arid areas or tropical environments but the mountain dragon occurs as far south as Tasmania and has the title of being the world's most southerly occurring dragon. In alpine areas on the mainland, the mountain dragon will remain dormant beneath snow in a self-constructed burrow.

Management: Surface rocks, fallen timber and dead trees, and exotic animals.

David Hunter

Skinks

Family Scincidae

Skinks are a cosmopolitan group of lizards that occur on most continents and occupy a broad range of environments. Many skinks are difficult to differentiate from each other, and in some cases inspection of scale arrangements are necessary to identify particular species. However, they can generally be distinguished from other lizard families in Australia by their shiny body scales. Most skinks are able to drop their tails when seized by predators. The tail, which usually regenerates within a few months, will often appear a different colour to the body. Australia has over 370 species of skinks, belonging to approximately 36 genera. The most species rich groups include: *Carlia*, *Cryptoblepharus*, *Ctenotus*, *Egernia*, *Eulamprus* and *Lerista*. All of these groups are represented in the Murray catchment. New skink species are being described in Australia each year, and in 2008 alone, more than a dozen new *Cryptoblepharus* species were described. Thirty-eight skink species occur in the Murray catchment.

Skinks comprise over 47% of the reptile fauna in the Murray catchment and it is not uncommon to find more than six species inhabiting the same area. In areas with suitable habitat, the number of co-occurring species can be higher. For example, 10 species have been recorded from Nail Can Hill Flora and Fauna Reserve in Albury and eight species have been recorded from an 80 ha woodland remnant near Koraleigh. Skinks within the Murray catchment can be divided into two groups depending on their distribution and hence, tolerance of different climatic conditions and temperatures. Approximately 58% of skinks are found in mesic environments, predominantly east of the Hume Highway. The remaining 42% are found in xeric environments, predominantly west of the Hume Highway. Terrestrial species account for over 60% of the skink fauna in the Murray catchment. These can be further divided into those that burrow in sandy soils, those associated with rocky areas and those that utilise leaf litter and grassy habitats. Five species are fossorial (soil-dwelling) and eight species are adapted to climbing and are associated with rocky outcrops or trees.

Eastern three-lined skink

Acritoscincus duperreyi

Damian Michael

Identifying features: Body colour brown with thin, dark, vertebral stripe. A broad, lateral stripe is bordered above and below by thin, white stripes. There is often a red to orange flush to the side of the neck which is more prominent in breeding males.

Snout–vent length: 80 mm.

Similar species: The red-throated skink lacks a vertebral stripe and pale stripes and species belonging to the genus *Pseudemoia* have paired scales on the top of the head.

Life-form: Terrestrial.

Reproductive mode: Oviparous.

Activity: Diurnal.

Distribution: Localised. Predominantly found in Kosciuszko National Park and also near Jingellic.

Shelter-sites: Fallen timber, surface rocks and grass tussocks.

Vegetation type: **Alpine woodland**, **montane forest**, wet forest and dry forest.

Conservation status: Common.

Additional notes: The eastern three-lined skink is unusual among alpine skinks in that it lays eggs.

Facts: For many reptiles, the gender of the offspring is not determined by genes but instead by internal nest temperature. One outcome is that at lower temperatures, males are produced, while at warmer temperatures females are produced. For egg-laying species that occur over a range of elevations it is conceivable that nests at higher sites will produce all male populations and nests at lower sites will produce all female populations. The eastern three-lined skink has overcome this problem by manipulating the amount of yolk each egg receives before the eggs are laid. Researchers have discovered that the size of the egg can override temperature in determining offspring sex in this species. Larger eggs

contain more yolk and produce female offspring whereas smaller eggs with less yolk produce males.

Management: Native grass, leaf litter and fine woody debris, and fallen timber and dead trees.

Damian Michael

Scincidae

Red-throated skink

Acritoscincus platynotum

Damian Michael

Identifying features: Body colour grey to brown with a broad, black lateral stripe along each side. The red-throated skink has a red throat which is more prominent during the breeding season.

Snout–vent length: 80 mm.

Similar species The eastern three-lined skink has a vertebral stripe and pale stripes above and below a broad, black, lateral stripe. Boulenger's skink has a pale stripe below the black lateral stripe.

Life-form: Terrestrial.

Reproductive mode: Oviparous.

Activity: Diurnal.

Distribution: Localised. Found east of Woomargama National Park.

Shelter-sites: Fallen timber, surface rocks and grass tussocks.

Vegetation type: Montane forest, wet forest and **dry forest**.

Conservation status: Common.

Additional notes: The red-throated skink is unusual among alpine skinks in that it lays eggs.

Facts: Communal nesting involves the clustering of eggs in the same location by several females of the same species or by females from multiple species. The phenomenon is widespread in many reptile families but is most common among higher elevation species including the red-throated skink. As many as 27 eggs of this species have been found clustered together beneath the one rock and the eggs from other species have also been found in the same nest. The reason why some species nest together may be more than just a response to limited egg-laying sites but may actually result in the improved fitness of the offspring. Researchers have found lizards that hatch from communal nests are larger and can move faster than lizards that hatch from

nests that contain a single clutch of eggs. This is due to a more stable microclimate within communal nests.

Management: Native grass, leaf litter and fine woody debris, and fallen timber and dead trees.

Damian Michael

Scincidae

McCoy's skink, highlands forest skink

Anepischtos maccoyi (formerly *Nannoscincus maccoyi*)

Peter Robertson

Identifying features: Body colour coppery-brown with pale flecks along the body and narrow, dark, dorsolateral stripes. The under belly is cream to yellow.

Snout–vent length: 50 mm.

Similar species: Coventry's skink has a broad, dark, side stripe and well-developed limbs.

Life-form: Fossorial.

Reproductive mode: Oviparous.

Activity: Nocturnal.

Distribution: Restricted. Found predominantly in Kosciuszko National Park.

Shelter-sites: Fallen timber and surface rocks.

Vegetation type: Alpine woodland, **montane forest**, **wet forest** and dry forest.

Conservation status: Uncommon.

Additional notes: McCoy's skink is sometimes referred to as the highlands forest skink because of its preference for cool moist environments. It is unusual among High Country lizards in that it lays clutches of eggs rather than giving birth to live young like many other cool-climate species. Females will often lay their eggs in a communal location and as many as 52 eggs have been found in the one nesting site.

Facts: McCoy's skink is atypical among skinks in that it has a preferred body temperature as low as 21°C making it potentially threatened to climate change. It tends to avoid sunny gaps in the forest canopy, instead preferring to burrow beneath the soil in cool, shaded areas. It is known to have one of the highest rates of water loss in any Australian lizard and is considered to be comparatively similar in its physiology to North American salamanders.

Management: Soil structure and integrity, surface rocks, leaf litter and fine woody debris, and fallen timber and dead trees.

Peter Robertson

Scincidae

Southern rainbow skink, four-fingered skink

Carlia tetradactyla

Damian Michael

Identifying features: Body colour grey to dark brown with a series of pale flecks along the back and onto the tail. Breeding males have an aqua lateral stripe bordered above and below by orange stripes. It is the only medium-sized skink with four fingers and five toes.

Snout–vent length: 64 mm.

Similar species: Grey's skink has four fingers although it much smaller. That species also has thin, black and pale lateral stripes.

Life-form: Terrestrial.

Reproductive mode: Oviparous.

Activity: Diurnal.

Distribution: Localised. Predominantly confined to the SWS bioregion.

Shelter-sites: Grass tussocks, leaf litter, fallen timber and surface rocks.

Vegetation type: Dry forest and **grassy box woodland**.

Conservation status: Common.

Additional notes: The southern rainbow skink can be relatively common in agricultural areas and responds well to grazing regimes that allow native grass species to mature. It also readily colonises tree plantings.

Facts: Most small species of lizards spend a portion of the day actively searching for food and at the same time avoid being eaten. The southern rainbow skink has developed a unique foraging technique that involves waving its tail from side to side and above its body. This behaviour is called caudal luring. It may serve to distract predators and draw attention to a skink's tail (which can grow again) rather than its head. It is not uncommon to encounter a southern rainbow skink

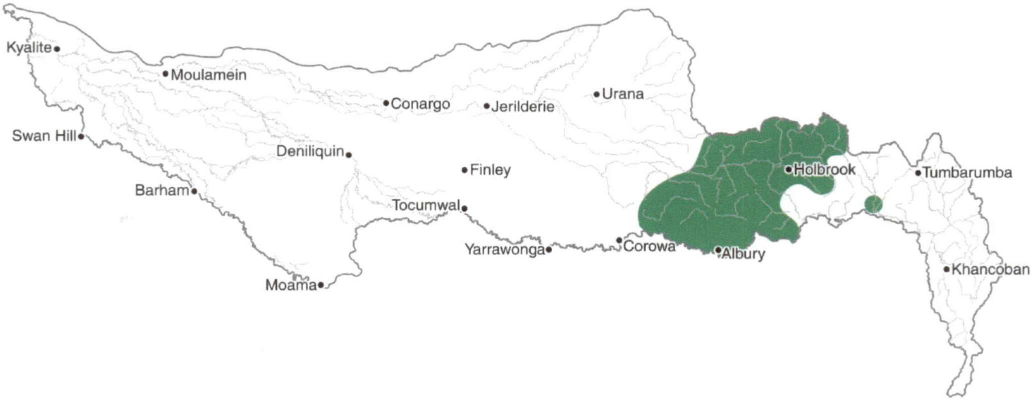

without a tail or with a partly regenerating stub. Tail loss can have consequences during the breeding season as an incomplete tail can impair breeding status and reduce the likelihood of acquiring a mate.

Management: Native grass, tree plantings, leaf litter and fine woody debris, and fallen timber and dead trees.

Damian Michael

Ragged snake-eyed skink

Cryptoblepharus pannosus (formerly *Cryptoblepharus carnabyi*)

Damian Michael

Identifying features: Body colour light to dark grey with a series of pale and dark flecks over the back and onto the tail. The limbs and digits are elongated and the body is flattened.

Snout–vent length: 40 mm.

Similar species: None.

Life-form: Arboreal and saxicolous.

Reproductive mode: Oviparous.

Activity: Diurnal.

Distribution: Widespread.

Shelter-sites: Bark, crevices of dead trees, fallen timber and rock outcrops.

Vegetation type: Dry forest, **grassy box woodland**, **black box woodland**, **sandhill woodland**, **red gum forest**, boree woodland and mallee woodland.

Conservation status: Common. One of the most abundant reptile species found in the Murray catchment.

Additional notes: The ragged snake-eyed skink is often encountered basking on the surfaces of dead trees and fallen timber and appears to have adapted well to living on human dwellings. Large numbers of ragged snake-eyed skinks can be found basking on the walls of buildings during warm weather. In areas with high amounts of fallen timber, densities may exceed one hundred individuals per hectare. Because of the skink's association with dead wood, it is likely that many individuals are transported to different areas within the Murray catchment.

Facts: The ragged snake-eyed skink is often found living in large aggregations. With so many lizards in one area, finding enough food can be difficult and may explain some unusual behaviour such as prey piracy. Similar members of this genus of lizards have been known to sit near ant trails and steal food items that are being carried by the ants back to their nests.

Management: Rocky outcrops, fallen timber and dead trees, and large mature trees.

Damian Michael

Brown-blazed wedgesnout skink
Ctenotus allotropis

David Webb

Identifying features: Body colour golden-brown with black lateral area enclosing pale spots, bordered above by white dorsolateral stripe. This species of *Ctenotus* lacks a vertebral stripe.

Snout–vent length: 55 mm.

Similar species: None, all other species of *Ctenotus* in the catchment have black vertebral stripes.

Life-form: Terrestrial.

Reproductive mode: Oviparous.

Activity: Diurnal.

Distribution: Localised. First recorded in the Murray catchment in 2006 in Buckingbong State Forest and Allen Carroll Flora and Fauna Reserve at Boree Creek.

Shelter-sites: Fallen timber, leaf litter and burrows.

Vegetation type: Grassy box woodland and sandhill woodland.

Conservation status: Rare.

Additional notes: The brown-blazed wedgesnout skink is one of a suite of species that has only recently been recorded in the Murray catchment because of an increasing number of surveys conducted there. It is an alert and swift-moving species and is likely to occur in similar habitat in the north of the catchment.

Facts: Some closely related lizards have extremely similar body shapes. Slight variations in the length of different body parts can reveal interesting insights into the evolution of different species. Within the *Ctenotus* lizards, there is a strong relationship between body size and the length of the hind limb. The larger species tend to have shorter back legs than the smaller

species. This relationship appears to have an influence on foraging behaviour, as smaller *Ctenotus* skinks, such as the brown-blazed wedgesnout skink, forage in open areas and have higher sprint speeds than larger species which tend to forage within dense vegetation.

Management: Leaf litter and fine woody debris, and fallen timber and dead trees.

David Webb

Scincidae

Murray striped skink, short-clawed skink

Ctenotus brachyonyx

David De Angelis

Identifying features: Body colour brown to grey with a black vertebral stripe not extending beyond back legs. Lateral area lacks spots.

Snout–vent length: 83 mm.

Similar species: The large striped skink has a vertebral stripe extending along the tail and lateral spots.

Life-form: Terrestrial.

Reproductive mode: Oviparous.

Activity: Diurnal.

Distribution: Restricted. First recorded in the Murray catchment in 2003 and known only from a single population near Koraleigh.

Shelter-sites: Spinifex tussocks and burrows.

Vegetation type: Mallee woodland.

Conservation status: Rare.

Additional notes: The Murray striped skink is one of the most range-restricted reptiles in the Murray catchment and is currently only known from a single sandy ridge near Koraleigh - an area which comprises only a few hectares of long-unburnt spinifex.

Facts: Spinifex grass is an extremely important habitat and can support a high diversity of species that are found nowhere else. The close association with spinifex and fire history can give rise to different reptile communities. Studies have shown that some species are more common in regenerating spinifex but long-unburnt spinifex patches contain the highest numbers of species. Maximum reptile diversity is reached when a mosaic of different age classes are maintained, thereby creating a range of conditions suitable for many different species.

Management: Native grass (including spinifex).

Kyalite
Moulamein
Swan Hill
Conargo
Jerilderie
Urana
Deniliquin
Finley
Holbrook
Tumbarumba
Barham
Tocumwal
Yarrawonga
Corowa
Albury
Khancoban
Moama

Spotted-back skink

Ctenotus orientalis

Damian Michael

Identifying features: Body colour pale brown to dark brown with a pale-edged vertebral stripe and numerous spot along the back and sides.

Snout–vent length: 82 mm.

Similar species: The large striped skink has a vertebral stripe extending along the tail. The royal skink lacks spots along the back.

Life-form: Terrestrial.

Reproductive mode: Oviparous.

Activity: Diurnal.

Distribution: Localised and patchily distributed across the Murray catchment.

Shelter-sites: Fallen timber, surface rocks and burrows.

Vegetation type: Grassy box woodland, sandhill woodland, **boree woodland** and **chenopod shrubland**.

Conservation status: Uncommon, although potentially common in localised areas.

Additional notes: The spotted-back skink excavates shallow, single chamber burrows often beneath clumping vegetation such as the scented mat-rush *Lomandra effusa*.

Facts: There is a growing amount of Australian literature on reptiles that are being recorded from new areas as well as outside their known geographical ranges. Such range extensions are usually a result of increased survey effort. The spotted-back skink is one species which had gone largely undetected in the Murray catch-

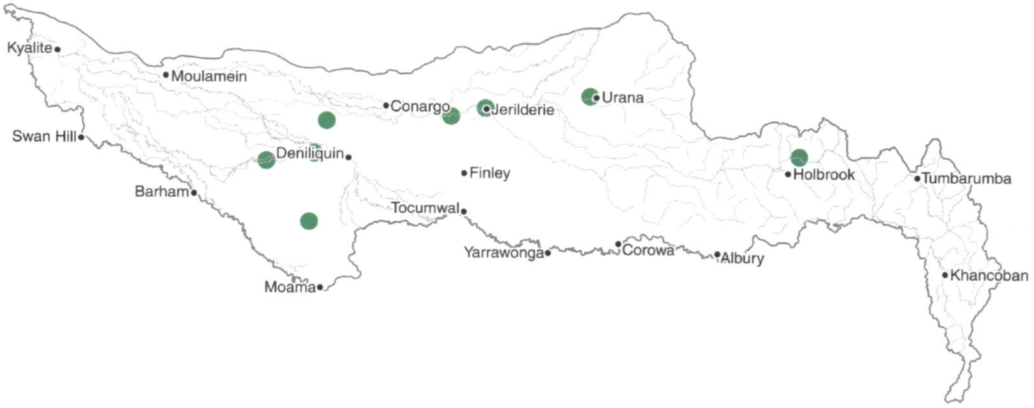

ment until recent surveys revealed it to be widely distributed in a range of habitats, particularly across the western half of the region. Like other *Ctenotus* skinks, it is extremely agile and when disturbed, it can retreat at lightening speed to a burrow without any hope of being caught. This may explain why the species went undetected for so many years in the Murray catchment.

Management: Soil structure and integrity, surface rocks, fallen timber and dead trees, and shrubs.

Damian Michael

Scincidae

Royal skink, regal striped skink

Ctenotus regius

Identifying features: Body colour brown with a narrow, pale-edged, dark vertebral stripe and a series of pale lateral spots bordered below by a thin, white stripe.

Snout–vent length: 73 mm.

Similar species: The large striped skink has a vertebral stripe extending along the tail. The spotted-back skink has spots either side of the vertebral stripe.

Life-form: Terrestrial.

Reproductive mode: Oviparous.

Activity: Diurnal.

Distribution: Restricted. First recorded in 2003 in the Murray catchment and known only from two locations in the far western part of the catchment.

Shelter-sites: Spinifex tussocks, grass tussocks, shrubs and burrows.

Vegetation type: Sandhill woodland and **mallee woodland**.

Conservation status: Rare.

Additional notes: The royal skink excavates a shallow burrow beneath vegetation, particularly spinifex tussocks. It is a common species across its geographical range in Australia.

Facts: The *Ctenotus* lizards are the most diverse reptile genus in Australia. Almost 100 species are currently recognised. In some parts of Australia, over a dozen different species co-occur in the one area. Studies have found that *Ctenotus* lizards avoid competing with each other by having different habitat and food requirements as well as maintaining different activity patterns. Some species forage in open spaces between plants, whereas others feed within dense vegetation. Some are active in

the late afternoon and some are active in the morning, and different prey is eaten by different species depending on their size.

Management: Soil structure and integrity, native grass, fallen timber and dead trees, and shrubs.

Large striped skink, robust ctenotus

Ctenotus robustus

Damian Michael

Identifying features: Body colour olive-brown with a broad, dark vertebral stripe that extends along tail and a series of pale lateral spots.

Snout–vent length: 123 mm.

Similar species: The Murray striped skink lacks spots along the side. The spotted-back skink has pale spots along the back. The royal skink has a vertebral stripe that does not extend onto tail and the bronze-blazed wedgesnout lacks a vertebral stripe.

Life-form: Terrestrial.

Reproductive mode: Oviparous.

Activity: Diurnal.

Distribution: Widespread except for the far eastern part of the Murray catchment.

Shelter-sites: Fallen timber, rock crevices, surface rocks and leaf litter.

Vegetation type: Dry forest, **grassy box woodland**, **sandhill woodland**, black box woodland and river red gum forest.

Conservation status: Common.

Additional notes: The large striped skink excavates a shallow burrow beneath surface rocks and fallen timber which terminates in a cavity large enough to accommodate a single individual. During the winter months, this species blocks off the entrance to the burrow and curls up with its head resting on the base of its tail. It is not uncommon to find these lizards lying motionless and covered in dew beneath their shelter-sites during the winter months.

Facts: Australian reptiles show a great diversity in the timing of their reproduction. Species in temperate regions usually breed during the spring and summer months but some species, especially those

that live in the Alps, breed during the autumn. Females of these species store sperm until the spring when embryonic fertilisation occurs. In tropical regions, some species breed during the wet season and others breed during the dry season. In regions that experience aseasonal rainfall, many species breed all year round.

Management: Native grass, rocky outcrops, surface rocks, and fallen timber and dead trees.

Damian Michael

Copper-tailed skink

Ctenotus taeniolatus

Damian Michael

Identifying features: Body colour tan-brown with a white edged, black vertebral stripe and three alternating black and white lateral stripes. There are no spots along the body.

Snout–vent length: 80 mm.

Similar species: The large striped skink has spots along the side of the body.

Life-form: Terrestrial.

Reproductive mode: Oviparous.

Activity: Diurnal.

Distribution: Localised. Predominantly found east of the Hume Highway between Albury and Holbrook.

Shelter-sites: Surface rock, fallen timber and self-made burrows.

Vegetation type: Dry forest and grassy box woodland.

Conservation status: Common.

Additional notes: The copper-tailed skink is capable of excavating a shallow burrow. However, in the Murray catchment it is also found sheltering between granite rock exfoliations. It constructs a small cavity from the sandy soil that accumulates between the rock layers and seals off the entrance during winter months.

Facts: All species of *Ctenotus* lay eggs and have clutches that range in size from one to nine eggs, depending on the species. A common relationship that exists for most egg-laying reptiles is that the number of eggs per clutch is related to the overall body size of the female. This is true for

individuals of a particular species as well as among different species. Larger species have larger clutches than smaller species and females lay more eggs as they mature. All *Ctenotus* species lay a single clutch each year. Other types of lizards can have up to three clutches each year, although other reptiles (particularly some snakes) will reproduce every second or third year.

Management: Rocky outcrops and surface rocks.

Damian Michael

Scincidae

Alpine she-oak skink

Cyclodomorphus praealtus

Nick Clemann

Identifying features: Body colour tan-brown to olive-green with a series of dark-edged scales forming lines along the back. The body is elongated and limbs short.

Snout–vent length: 119 mm.

Similar species: None.

Life-form: Terrestrial.

Reproductive mode: Viviparous.

Activity: Diurnal and crepuscular.

Distribution: Restricted to elevations above 1500 m in Kosciuszko National Park.

Shelter-sites: Tussock grass, fallen timber and surface rocks.

Vegetation type: Alpine grassland.

Conservation status: Rare.

Additional notes: The alpine she-oak skink gives birth to between two and nine young during mid-simmer and is unusual among reptiles in that it defends the young from predators. It is often seen basking on tussock grass and will quickly disappear when disturbed. This species appears to move like a snake.

Facts: Alpine species, such as the alpine she-oak skink, are part of a group of reptiles that may be at risk of extinction due to the combined influence of habitat modification and climate change. Computer models indicate that increases in temperature of as little as 0.5°C can have a significant impact on alpine fauna by reducing the extent and duration of snow cover. With a warmer climate, alpine plants may be replaced by other kinds of vegetation

and entire plant communities might disappear. This may have severe consequences for High Country reptiles with limited distributions as they will be unable to migrate to other areas of suitable habitat. In some cases their preferred habitat may even cease to exist.

Management: Native grass, surface rocks, and fallen timber and dead trees.

Scincidae

Cunningham's skink

Egernia cunninghami

Damian Michael

Identifying features: Body colour dark brown to black with a series of lighter markings and spots along the back. The body scales are rough and tail spiny.

Snout–vent length: 200 mm.

Similar species: The much smaller tree crevice skink and black rock skink have smoother scales along the body and tail.

Life-form: Saxicolous and semi-arboreal.

Reproductive mode: Viviparous.

Activity: Diurnal.

Distribution: Localised and patchily distributed in the Murray catchment as far west as Albury.

Shelter-sites: Rock crevices, tree stumps and hollow logs.

Vegetation type: Dry forest and grassy box woodland.

Conservation status: Uncommon, although locally abundant in some areas.

Additional notes: Cunningham's skink is a social lizard that inhabits rocky outcrops. It is often found in closely related, extended family groups, with as many as 20 individuals. This species includes plant material in its diet.

Facts: To maintain a viable population of any species, new individuals must constantly be recruited to balance the effects of mortality and juvenile dispersal. Some species, such as Cunningham's skink show high site fidelity. That is, they spend their entire lives in a closely related family group on a single rocky outcrop. Box sexes are known to disperse but these events are infrequent. Habitat fragmentation can disrupt dispersal. Thus, group sizes are generally larger in undisturbed environments. Having so many close relatives living in the

same area makes inbreeding avoidance important for maintaining a genetically variable population. Cunningham's skink avoids inbreeding by forming long-term mate bonds with the least related member of a group.

Management: Rocky outcrops, and fallen timber and dead trees.

Damian Michael

Black rock skink

Egernia saxatilis ssp. *intermedia*

Identifying features: Body colour uniformly dark grey to black with a series of darker flecks and lighter cross bands along each side of the body.

Snout–vent length: 135 mm.

Similar species: The tree crevice skink has smoother less keeled body scales and more prominent white upper lip scales. Cunningham's skink has a series of spines on the tail.

Life-form: Saxicolous.

Reproductive mode: Viviparous.

Activity: Diurnal.

Distribution: Localised. Predominantly found east of Jingellic in the Murray catchment.

Shelter-sites: Rock crevices and surface rocks.

Vegetation type: Alpine woodland, montane forest and **dry forest**.

Conservation status: Uncommon.

Additional notes: The black rock skink is a social lizard and forms 'nuclear'-type family groups consisting of an adult male, an adult female and one or more generations of juvenile offspring.

Facts: Many lizard species are solitary. However, most *Egernia* species are social and live in closely related family groups. Adult black rock skinks aggressively defend their home site in rocky outcrops and even attack unrelated juveniles of the same species. To prevent fatal encounters, juveniles of the black rock skink recognise the scent of related family members, and thus avoid contact with unfamiliar and potentially unfriendly adults. Being able to recognise closely related family members from

non-related individuals is thought to be an important factor in the evolution of social behaviour in lizards.

Management: Rocky outcrops, and fallen timber and dead trees.

Damian Michael

Tree crevice skink

Egernia striolata

Damian Michael

Identifying features: Body colour dark grey with scattered, pale flecks over the back and a broad, dark ragged-edged lateral band. The upper lips and throat is white.

Snout–vent length: 119 mm.

Similar species: The black rock skink is more robust, has rougher body scales, and lacks prominent white upper lips. Cunningham's skink has a series of spines on the tail.

Life-form: Saxicolous and arboreal.

Reproductive mode: Viviparous.

Activity: Diurnal.

Distribution: Localised. Predominantly found in the SWS bioregion but also occurs along the Murray River floodplain.

Shelter-sites: Rock crevices, tree stumps, exfoliating bark and crevices of fallen and standing dead trees.

Vegetation type: Dry forest, **grassy box woodland** and **river red gum forest**.

Conservation status: Common, although many lowland populations have declined in the SWS bioregion.

Additional notes: The tree crevice skink is often solitary but can form closely related family groups on granite outcrops and occasionally large, mature river red gum trees along the Murray River. It is often seen basking on vertical rock faces and will quickly disappear into a crevice when disturbed.

Facts: In a study in the SWS bioregion, the tree crevice skink was found to be more abundant on structurally complex rocky outcrops that contained lots of large granite boulders, and that were surrounded by native vegetation and tree plantings. Family groups were also more likely to occupy the largest rocks near the peak of the outcrops,

whereas immature lizards were more commonly encountered on smaller rocks near the edge of an outcrop.

Management: Rocky outcrops, tree plantings, fallen timber and dead trees, and large mature trees.

Damian Michael

Scincidae

Yellow-bellied water skink, Heatwoles's water skink

Eulamprus heatwolei

Damian Michael

Identifying features: Body colour brown with numerous black flecks along the back and a ragged-edged, broad lateral band. Belly is pale cream to bright yellow and dorsolateral stripe extends from the eye to the front limbs. The front section of the ear-opening is black.

Snout–vent length: 100 mm.

Similar species: The southern water skink lacks a dorsolateral stripe and has a white section to the front of ear-opening. The alpine water skink has prominent dorsolateral stripes.

Life-form: Terrestrial.

Reproductive mode: Viviparous.

Activity: Diurnal.

Distribution: Localised. Predominantly found east of Woomargama National Park but also occurs along the Murray River floodplain as far west as Mathoura.

Shelter-sites: Fallen timber and surface rocks.

Vegetation type: Wet forest, **dry forest**, grassy box woodland and **river red gum forest** and wetlands.

Conservation status: Common.

Additional notes: The yellow-bellied water skink is common around creeks in the eastern part of the catchment and penetrates further west than any other *Eulamprus* species.

Facts: Female lizards will try and maximise the genetic diversity of their offspring by mating with several partners. Males will also try to ensure that it is their genes that are passed on. In species that are

sexually receptive for a short period of time, such as the yellow-bellied water skink, males will guard their mates from other males. When rival males venture too close, a dominant male will chase them off and signal his intent by bobbing his head in annoyance. Recent studies have found mate guarding does not always work as more than two-thirds of copulations are extra-pair mating events.

Management: Wetlands, surface rocks, and fallen timber and dead trees.

Nick Clemann

Alpine water skink

*Eulamprus kosciuskoi**

David Hunter

Identifying features: Body colour dark brown with thin, black vertebral stripes, a broad, dark lateral band flecked with white spots, bordered above by yellow dorsolateral stripes.

Snout–vent length: 85 mm.

Similar species: The yellow-bellied water skink has a dorsolateral stripe extending from the eye to the front limbs. The southern water skink lacks dorsolateral stripes.

Life-form: Terrestrial.

Reproductive mode: Viviparous.

Activity: Diurnal.

Distribution: Restricted. Known only from Kosciuszko National Park.

Shelter-sites: Fallen timber and surface rocks.

Vegetation type: Alpine grassland and wetlands.

Conservation status: Uncommon.

Additional notes: The alpine water skink, as the name suggests, is common around the margins of streams, bogs and alpine meadows. It often catches aquatic prey from the surface of the water, although, like most lizards, it is an opportunistic predator that will eat anything it encounters, except the species own young.

Facts: Many lizard species are solitary and are aggressive toward other lizards of the same species. Of the three *Eulamprus* species that occur in the Murray catchment, the alpine water skink is the most aggressive, and will actively defend its favourite basking sites and foraging areas by chasing off intruders. When disturbed, a water skink

will quickly take cover or dive into a stream, but will return to the rock or log on which they were basking within a few minutes.

Management: Wetlands, surface rocks, and fallen timber and dead trees.

* The spelling of Mount Kosciuszko was changed to include the letter 'z', after the correct polish spelling. However, the rules which govern the naming of animals, dictates that the original 'incorrect' spelling is kept for the alpine water skink.

Nick Clemann

Scincidae

Southern water skink, highlands water skink

Eulamprus tympanum

Damian Michael

Identifying features: Body colour is golden-brown with dark flecks along the back and broad, black lateral stripes enclosing small pale spots. Front of ear opening is pale.

Snout–vent length: 93 mm.

Similar species: The yellow-bellied water skink has a dorsolateral stripe extending from the eye to the front limbs and the front of the ear opening is black. The alpine water skink has prominent dorsolateral stripes.

Life-form: Terrestrial.

Reproductive mode: Viviparous.

Activity: Diurnal.

Distribution: Localised. Predominantly found in Kosciuszko National Park.

Shelter-sites: Fallen timber and surface rocks.

Vegetation type: Alpine woodland, **montane forest**, **wet forest**, dry forest and wetlands.

Conservation status: Common.

Additional notes: The southern water skink gives birth to between two and six young. It feeds on a range of arthropods and is known to eat small lizards.

Facts: The three *Eulamprus* species that occur in the Murray catchment are similar in appearance and occupy similar environments, such as stream sides and moist areas. Often, two species will overlap in range. However, the yellow-bellied water skink occurs at low altitudes and prefers northerly, more open, rocky environments. The southern water skink occurs in mid-zone elevations and the alpine water skink occurs at high elevations, including the summit of Mount Kosciuszko. It is

therefore unlikely that all three species co-occur in the same area.

Management: Wetlands, surface rocks, and fallen timber and dead trees.

Nick Clemann

Scincidae

Three-toed skink

Hemiergis decresiensis ssp. *talbingoensis*

Identifying features: Body colour pale grey to chocolate-brown with a series of thin stripes along the body. Belly is cream to bright yellow and is the only species with three fingers and toes.

Snout–vent length: 60 mm.

Similar species: The wood-mulch slider has a dark lateral stripe and lacks a yellow belly and McCoy's skink has five fingers and toes.

Life-form: Fossorial.

Reproductive mode: Viviparous.

Activity: Diurnal.

Distribution: Localised. Found predominantly east of the Hume Highway.

Shelter-sites: Fallen timber, surface rocks and leaf litter.

Vegetation type: Alpine woodland, **montane forest**, **wet forest**, dry forest and **grassy box woodland**.

Conservation status: Common.

Additional notes: The three-toed skink is often abundant where it is moist. During summer months they become scarce and are difficult to detect, and probably spend most of the time deep within the soil layer or beneath mats of leaf litter. This species is often found co-inhabiting with small black ants, which are known to be its preferred food.

Facts: As the name suggests, the three-toed skink has a reduced number of digits. Many species that burrow in loose soil and leaf litter have elongated bodies, a trait that enables them to move through friable soil with apparent ease. When disturbed, the

three-toed skink will quickly burrow into the soil. Its bright yellow belly may serve to startle predators.

Management: Soil structure and integrity, leaf litter and fine woody debris, and fallen timber and dead trees.

Damian Michael

Scincidae

Delicate skink, grass skink

Lampropholis delicata

Damian Michael

Identifying features: Body colour grey-brown to dark brown with black lateral. Head is coppery-brown. Pale dorsolateral stripe in some individuals.

Snout–vent length: 51 mm.

Similar species: The garden skink has pale flecks along the back and usually a dark vertebral stripe. Coventry's skink is larger and has faint orange dorsolateral stripes.

Life-form: Terrestrial.

Reproductive mode: Oviparous.

Activity: Diurnal.

Distribution: Localised. Found east of Woomargama National Park in the Murray catchment.

Shelter-sites: Fallen timber, surface rocks and leaf litter.

Vegetation type: Montane forest and dry forest.

Conservation status: Common.

Additional notes: The delicate skink is a common species in south-eastern Australia and is often heard scurrying into leaf litter. Between two and five eggs are laid per clutch, often in communal nesting sites.

Facts: During the breeding season, many male skinks acquire breeding colours that distinguish them from adult females and non-breeding males. This is called sexual dichromatism. These breeding hues last for a short period but a second type of dichromatism is more permanent. In some species, such as the delicate skink, females possess a fine white stripe along each side of the body. Females from Queensland populations tend to be strongly dichro-

matic, whereas within southern Australian populations, only a small percentage of females have white markings.

Management: Native grass, leaf litter and fine woody debris, and fallen timber and dead trees.

Damian Michael

Scincidae

Garden skink, grass skink

Lampropholis guichenoti

Damian Michael

Identifying features: Body colour grey-brown usually with a dark vertebral stripe and pale flecks along the back. A dark lateral stripe is bordered above and below by thin, white stripes.

Snout–vent length: 48 mm.

Similar species: The delicate skink lacks a vertebral stripe and has dark flecks along the back.

Life-form: Terrestrial.

Reproductive mode: Oviparous.

Activity: Diurnal.

Distribution: Localised. Predominantly found east of the Yambla Range in the Murray catchment but also occurs in river red gum forest along the Murray River floodplain.

Shelter-sites: Fallen timber, surface rocks and leaf litter.

Vegetation type: Alpine woodland, **montane forest**, **dry forest**, grassy box woodland and **river red gum forest**.

Conservation status: Common.

Additional notes: As the name suggests, the garden skink is a common species in urban areas of south-eastern Australia. During the breeding season, males become aggressive towards other males. Often, mating 'balls' occur, which involves a male and female in copulation and several additional males trying to separate them. Challenging males will bite each other and attempt to separate the mating pair.

Facts: Many reptiles become dull in colour as they get older, although others like some species of dragons and geckos are capable of short-term colour changes. However, lizards in the skink family are generally not able to change their body colour over short

periods of time. Young individuals often support brightly coloured tails and bold contrasting body patterns and spots but these markings fade as they mature. The garden skink is unusual in that its body colour appears to lighten after exposure to high temperatures and darkens when it is inactive during cool conditions.

Management: Native grass, leaf litter and fine woody debris, and fallen timber and dead trees.

Damian Michael

Scincidae

Bougainville's skink, southern lerista, south-eastern slider
Lerista bougainvillii

Damian Michael

Identifying features: Body colour grey to brown with a broad, black lateral stripe and an orange to brown cylindrical tail. Body is elongated and limbs reduced.

Snout–vent length: 70 mm.

Similar species: Boulenger's skink has well developed limbs and a white lateral stripe.

Life-form: Fossorial.

Reproductive mode: Oviparous.

Activity: Diurnal.

Distribution: Widespread but patchily distributed in the western half of the Murray catchment.

Shelter-sites: Surface rocks, fallen timber, leaf litter and other ground debris.

Vegetation type: Dry forest, grassy box woodland, **sandhill woodland**, chenopod shrubland and mallee woodland.

Conservation status: Uncommon, although potentially locally abundant in some areas.

Additional notes: Bougainville's skink has a patchy distribution in the Murray catchment but is difficult to detect because it shelters beneath leaf litter and soil. Recent studies in the Murray catchment have found this species uses roof tiles placed in tree plantings, and often occurs in the nests of small black ants.

Facts: There are two modes of reproduction in Australian reptiles, egg-laying and giving birth to live young. Live-bearing species are more likely to occur at higher elevations and cooler temperatures. Most reptiles employ one method of reproduction, but Bougainville's skink is unusual as it contains both egg-laying and live-bearing populations. In Tasmania and islands in Bass Strait, populations are entirely live-bearing, whereas those on the mainland lay eggs. Mainland populations near the

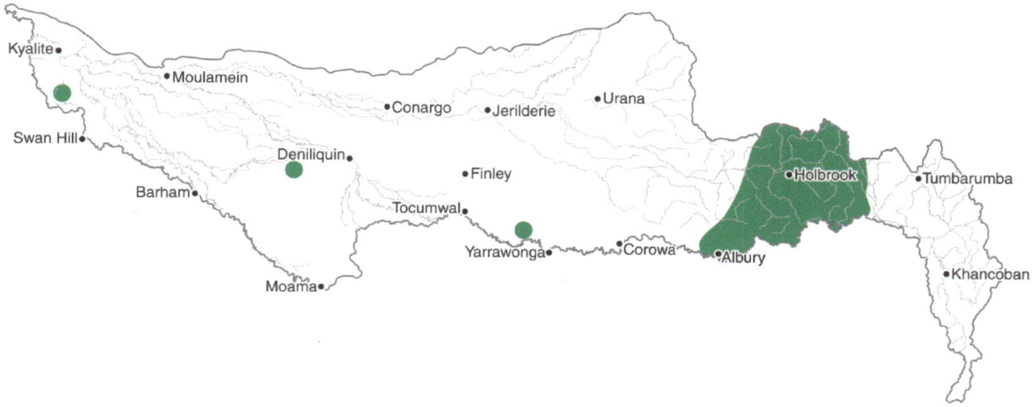

south coast retain their eggs for a longer period than northern populations and their eggs hatch within a couple weeks of being laid.

Management: Soil structure and integrity, tree plantings, leaf litter and fine woody debris, surface rocks, and fallen timber and dead trees.

Damian Michael

Eastern robust slider, spotted burrowing skink, spotted lerista
Lerista punctatovittata

David Webb

Identifying features: Body colour fawn or tan-brown with lines of black spots along the body and onto the tail. The only species with one finger (sometimes two), two toes and tiny limbs.

Snout–vent length: 100 mm.

Similar species: None.

Life-form: Fossorial.

Reproductive mode: Oviparous.

Activity: Diurnal.

Distribution: Restricted. First recorded in the Murray catchment in 2003 and known only from four locations in the far western part of the region.

Shelter-sites: Fallen timber, leaf litter and sandy soils.

Vegetation type: Mallee woodland.

Conservation status: Rare.

Additional notes: The eastern robust slider produces between two and four eggs per clutch and is often found in small aggregations. It is difficult to detect without the aid of trapping equipment (such as pitfall traps), although it is often encountered after heavy rain.

Facts: Unlike mammals, which primarily walk on legs positioned beneath their bodies, snakes and lizards have an undulating gait due to their lower centre of gravity and, in the case of lizards, outwardly positioned legs. Being low to the ground means they have to bend their spine and torso sideways to swing their legs in front and propel themselves forward. The lateral undulating 'S' curve is the most common form of locomotion used by most reptiles. Snakes have perfected this form of locomotion, but legless lizards and limbless skinks such as the spotted burrowing skink also use this method. By pushing their bodies

against a rough surface they can move with surprising speed and have the appearance of swimming through sand.

Management: Soil structure and integrity, leaf litter and fine woody debris, and fallen timber and dead trees.

David Webb

Scincidae

Wood-mulch slider, three-toed lerista

Lerista timida (formerly *Lerista muelleri*)

Identifying features: Body colour grey to brown with a series of fine, dark, broken lines along the body. Dark lateral stripe and three fingers and toes. Juveniles often have a red flush to the tail.

Snout–vent length: 50 mm.

Similar species: The three-toed skink has a yellow belly and no white lateral stripe.

Life-form: Fossorial.

Reproductive mode: Oviparous.

Activity: Diurnal.

Distribution: Widespread, especially in the western part of the Murray catchment.

Shelter-sites: Loose soil, leaf litter, fallen timber and other ground debris.

Vegetation type: Grassy box woodland, **sandhill woodland**, black box woodland and **mallee woodland**.

Conservation status: Common.

Additional notes: The wood-mulch slider produces two eggs per clutch and is often found within and beneath rotting timber. It has a patchy distribution in the Murray catchment and may include several undescribed subspecies.

Facts: There are over 80 *Lerista* species in Australia making them the second largest genus after *Ctenotus* (which contains almost 100 species). They exhibit the widest range of limb conditions and arrangements of any other group of lizard. Species range from having fully formed limbs with five digits to no front limbs and one or two digits on the hind limb. They can also be completely limbless, a condition that is even more reduced than primitive snakes such as pythons, which still retain remnant pelvic bones and hind limb spurs. Species with well-developed limbs and a complete number of digits tend to be surface-active during the day. Species with no limbs or reduced digits tend to be active on

dusk or during the night and 'swim' beneath sandy soil. It is not uncommon to find over a dozen *Lerista* species occupying the one environment in desert regions of Australia.

Management: Soil structure and integrity, leaf litter and fine woody debris, and fallen timber and dead trees.

Scincidae

Snowy Mountains rock skink

*Liopholis guthega (*formerly *Egernia guthega)*

Nick Clemann

Identifying features: Body colour blue to grey with three pale vertebral stripes enclosing a series of white spots.

Snout–vent length: 111 mm.

Similar species: White's skink is lighter and has distinctive black-edged white spots along each side.

Life-form: Terrestrial and saxicolous.

Reproductive mode: Viviparous.

Activity: Diurnal.

Distribution: Restricted to above 1600 m elevation in Kosciuszko National Park.

Shelter-sites: Burrows, fallen timber and surface rocks.

Vegetation type: Alpine grassland and alpine woodland.

Conservation status: Uncommon.

Additional notes: The Snowy Mountains rock skink is restricted to high elevation granite outcrops in the Snowy Mountains and Victorian Bogong High Plains. It has been recorded from elevations of 1940 m making it one of Australia's highest occurring lizard species. It is one of a group of high-elevation species that may be affected by increased temperature associated with climate change in the years to come. The Snowy Mountains rock skink lives in large colonial groups on exposed bolder-strewn hillsides and shelters within a network of burrows which it excavates at the base of boulders or beneath shrubs.

Facts: New species of reptiles are being discovered and described every year in Australia, even from relatively well-known parts of the country. Molecular analysis of the *Egernia* lizards in 2002 revealed the Snowy Mountain's rock skink to be a new

species which has recently been assigned a new genus name *Liopholis*.

Management: Rocky outcrops, surface rocks and shrubs.

Scincidae

Tan-backed rock skink

Liopholis montana (formerly *Egernia montana*)

David Hunter

Identifying features: Body colour tan-brown with a grey head and sides. A series of lighter spots run along each side of the body.

Snout–vent length: 111 mm.

Similar species: White's skink has distinctive black-edged white spots along each side. The Snowy Mountains rock skink has spots and stripes along the back.

Life-form: Terrestrial.

Reproductive mode: Viviparous.

Activity: Diurnal.

Distribution: Restricted. Found only in Kosciuszko National Park.

Shelter-sites: Burrows, fallen timber and surface rocks.

Vegetation type: Alpine woodland and montane forest.

Conservation status: Uncommon.

Additional notes: The tan-backed rock skink was formally described in 2002 and like other closely related skinks, lives in burrow systems beneath rock slabs and jumbled rock piles.

Facts: Scat piling is an unusual behaviour but it is common to many skinks that live in small colonies or family groups. These lizards defecate in the one spot, either outside the burrow entrance or on a prominent location on rock ledges. Scat piles may act as territorial markers but they may also serve another function. Insects are often attracted to these dung piles, which may provide these scat-piling skinks with an easy meal not far from cover.

Management: Rocky outcrops, surface rocks, fallen timber and dead trees, and shrubs.

Nick Clemann

Scincidae

White's skink

*Liopholis whitii (*formerly *Egernia whitii)*

Damian Michael

Identifying features: Body colour blue-grey to grey-brown usually with two dark stripes along the back enclosing a series of fine white spots. Along each side are black-edged white spots.

Snout–vent length: 113 mm.

Similar species: The Snowy Mountains rock skink lacks black-edged white spots along the side. The tan-backed rock skink has a broad, tan-coloured back and lacks spots.

Life-form: Terrestrial.

Reproductive mode: Viviparous.

Activity: Diurnal.

Distribution: Localised. Predominantly found in the eastern part of the Murray catchment.

Shelter-sites: Rock crevices and jumbled rock piles.

Vegetation type: Montane forest and dry forest.

Conservation status: Uncommon.

Additional notes: White's skink is a social lizard, living in closely related family groups. It shelters in burrows at the base of rock piles and boulders and is quick to retreat to cover when disturbed.

Facts: Being able to accurately identify a lizard species before it disappears under cover can be a challenge, even for experienced herpetologists. Often, size, body colour, unique markings and behaviour are the only features required to identify some species. Other identifying features, such as counting scales and the number of digits can be useful, but in both cases the animal needs to be caught, in which case a scientific licence and animal care and ethics approval must be gained from the relevant wildlife authorities. To compound the problem of identification, some lizard species, such as White's skink, can vary considerably in body pattern, even within the same population as well as across its geographical range. The most common form is described above but White's skink can occur as a plain-backed and weakly spotted form. Species which exhibit

diversity in body patterns are often a complex of subspecies.

Management: Rocky outcrops, surface rocks, and fallen timber and dead trees.

Grey's skink, dwarf skink

Menetia greyii

Identifying features: Body colour grey-brown with dark lateral stripe bordered below by a thin white stripe. Digits include four fingers and five toes. Breeding males have an orange throat and yellow belly.

Snout–vent length: 38 mm.

Similar species: Boulenger's skink is larger and has five fingers.

Life-form: Terrestrial.

Reproductive mode: Oviparous.

Activity: Diurnal.

Distribution: Widespread, occurring as far east as the Hume Highway.

Shelter-sites: Fallen timber, surface rocks, leaf litter and grass tussocks.

Vegetation type: Grassy box woodland, sandhill woodland, black box woodland, **boree woodland**, **chenopod shrubland** and mallee woodland.

Conservation status: Common.

Additional notes: Grey's skink produces up to three eggs per clutch and is a common and abundant lizard in grassy environments in the Murray catchment, although it is often overlooked due to its small size.

Facts: Most vertebrates reproduce sexually by males fertilising female egg cells. However, some species are able to reproduce without the need of sperm. That is, they reproduce asexually. Until recently, asexual reproduction was only known to occur in geckos and some snakes. Recent studies have found that Grey's skink is also capable of reproducing without the assistance of males. In some populations, the offspring are a replica of the parent. These lizards are neither male nor female and do not require a partner to reproduce. Genetic information is passed to the offspring via a single parent, resulting in a genetically identical clone, a process known as parthenogenesis.

Management: Native grass, leaf litter and fine woody debris, and fallen timber and dead trees.

Damian Michael

Scincidae

Saltbush skink, chenopod morethia, samphire skink

Morethia adelaidensis

Damian Michael

Identifying features: Body colour grey to brown with a series of dark flecked rows along the back. A ragged-edged white stripe along the sides is bordered above by a dark broken line. Breeding males have an orange flush to the side of the neck and the scales above the eyes are pointed.

Snout–vent length: 53 mm.

Similar species: Boulenger's skink has a straight-edged, white lateral stripe and square-shaped scales above the eyes.

Life-form: Terrestrial.

Reproductive mode: Oviparous.

Activity: Diurnal.

Distribution: Widespread, but patchily distributed in the western part of the region. First recorded in the Murray catchment in 2005.

Shelter-sites: Fallen timber, leaf litter and low shrubs.

Vegetation type: Boree woodland and **chenopod shrubland**.

Conservation status: Uncommon, although potentially abundant in areas supporting suitable habitat.

Additional notes: The saltbush skink produces up to six eggs per clutch, often in communal nesting sites. This species prefers more open habitats than the closely related Boulenger's skink, although both species may co-occur in boree woodland.

Facts: Unlike other groups of animals such as birds, distinguishing the sex of reptiles can be difficult. Both sexes often look similar. During the breeding season however, the males of many species display conspicuous orange or red flushes on the throat and sides of the neck. Throat colouration is most striking in *Morethia* skinks. Species such as the saltbush skink probably use these temporary colours to attract suitable mates. After the breeding season, throat colour

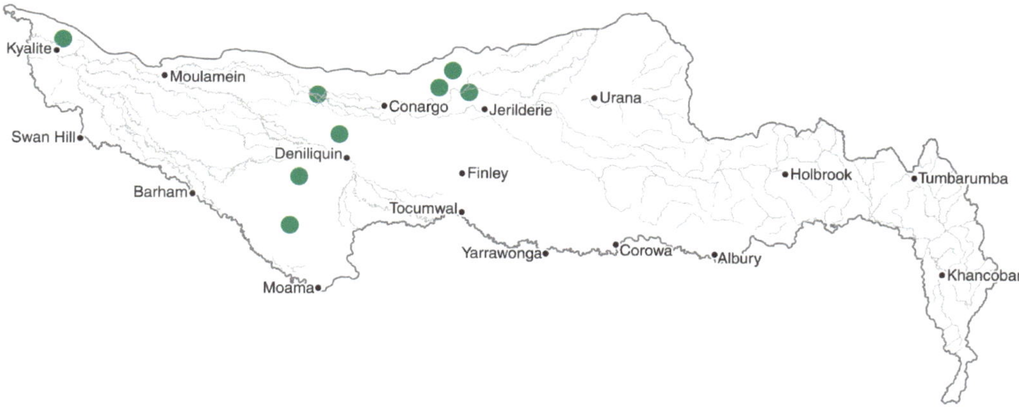

disappears and the only way to tell the sexes of mature species is based on body size. Generally, mature female skinks are larger and have shorter heads than males.

Management: Leaf litter and fine woody debris, fallen timber and dead trees, and shrubs.

Damian Michael

Boulenger's skink

Morethia boulengeri

Identifying features: Body colour grey-brown with a black lateral stripe bordered below by a white stripe. Breeding males have an orange throat and juveniles often have reddish tails. Scales above the eyes are square-shaped.

Snout–vent length: 50 mm.

Similar species: The saltbush skink has a ragged-edged white lateral stripe and pointed scales above the eye. Grey's skink has four fingers.

Life-form: Terrestrial.

Reproductive mode: Oviparous.

Activity: Diurnal.

Distribution: Widespread penetrating along the Murray River valley as far east as Jingellic.

Shelter-sites: Fallen timber, exfoliating bark of tree stumps, leaf litter, debris and surface rocks.

Vegetation type: Dry forest, **grassy box woodland**, **sandhill woodland**, **black box woodland**, **river red gum forest**, chenopod shrubland and mallee woodland.

Conservation status: Common. One of the most abundant reptile species found in the catchment.

Additional notes: Boulenger's skink is the most numerous reptile species in the Murray catchment. It is capable of reproducing three times per year. In areas with abundant fallen timber, the species reaches densities as high as several hundred per hectare.

Facts: Surveys have found Boulenger's skink inhabits most vegetation types in the Murray catchment, being absent only from montane and alpine vegetation types, where it is replaced by the similarly abundant garden skink. Boulenger's skink is usually terrestrial, however, this species is often encountered several metres above the ground sheltering behind exfoliating bark. It will use almost any type of cover for shelter, including dry cattle dung, livestock carcasses and a range of waste material.

Management: Native grass, rocky outcrops, tree plantings, leaf litter and fine woody debris, surface rocks, fallen timber and dead trees, shrubs, and large mature trees.

Damian Michael

Scincidae

Coventry's skink, southern forest skink

Niveoscincus coventryi

Damian Michael

Identifying features: Body colour coppery-brown with pale flecks and a broad, black lateral stripe bordered above by a narrow orange stripe.

Snout–vent length: 50 mm.

Similar species: McCoy's skink has reduced limbs and cream to yellow belly.

Life-form: Terrestrial.

Reproductive mode: Viviparous.

Activity: Diurnal.

Distribution: Localised. Found in the far eastern part of the Murray catchment.

Shelter-sites: Fallen timber and surface rock.

Vegetation type: Alpine woodland, **montane forest**, **wet forest** and dry forest.

Conservation status: Common.

Additional notes: Coventry's skink gives birth to up to seven young in late summer. It is an abundant skink in forest vegetation types and is often found near large decaying logs in forest clearings.

Facts: Body colour in skinks often reveals other information about lifestyle. Species that occur in mountainous areas and are surface-active are generally grey in colour while species that spend large amounts of time beneath debris are brown. Skinks that live in drier more open regions of Australia tend to have body colours that match the substrate. Those species which inhabit sandy regions are often reddish-brown. Camouflage is one reason why skinks from different areas differ in body colour. However, other factors such as protection from solar radiation and temperature regulation may also be important.

Management: Leaf litter and fine woody debris, and fallen timber and dead trees.

Kyalite
Moulamein
Conargo
Jerilderie
Urana
Swan Hill
Deniliquin
Finley
Holbrook
Tumbarumba
Barham
Tocumwal
Yarrawonga
Corowa
Albury
Khancoban
Moama

Woodland tussock skink, mountain log skink

Pseudemoia entrecasteauxii

David Hunter

Identifying features: Body colour brown with a thin, dark vertebral stripe and two pale, broken lateral stripes. Breeding males have an orange flush to the sides.

Snout–vent length: 60 mm.

Similar species: The southern grass skink and glossy grass skink have continuous pale lateral stripes and no small pale flecks on the back.

Life-form: Terrestrial.

Reproductive mode: Viviparous.

Activity: Diurnal.

Distribution: Localised. Found only in the far eastern part of the Murray catchment.

Shelter-sites: Fallen timber, ground debris and surface rocks.

Vegetation type: Alpine woodland, **montane forest**, wet forest and **dry forest**.

Conservation status: Uncommon.

Additional notes: The woodland tussock skink gives birth to up to seven young and is a common lizard in grassy areas with lots of fallen timber. The species is highly variable in colour pattern across its range.

Facts: Finding a suitable over-wintering site is not an issue for reptiles that occur in tropical parts of Australia. This is because minimum temperatures during the winter months are not low enough to affect daily activity. However, for those species which occur in cold and moist environments, such as the woodland tussock skink, finding a thermally-stable over-wintering site is paramount to their survival. Large aggregations of lizards have been found in communal aggregations deep within large rotting logs. This is probably a response to a limited number of suitable shelter-sites within a species' home range. Over-wintering sites are often shared by a number of different species.

Management: Native grass, leaf litter and fine woody debris, and fallen timber and dead trees.

Damian Michael

Southern grass skink, grassland tussock skink

Pseudemoia pagenstecheri

Nick Clemann

Identifying features: Body colour olive-brown with three dark vertebral stripes. Tan-brown lateral stripe is bordered above by a thin, white stripe and adult males have a thin, lower lateral orange stripe.

Snout–vent length: 60 mm.

Similar species: The woodland tussock skink has broken lateral stripes and the dorsolateral stripe on the glossy grass skink is positioned on the third scale row from the vertebral stripe.

Life-form: Terrestrial.

Reproductive mode: Viviparous.

Activity: Diurnal.

Distribution: Restricted. Known only from Kosciuszko National Park.

Shelter-sites: Grass tussocks, leaf litter and fallen timber.

Vegetation type: Alpine grassland.

Conservation status: Uncommon.

Additional notes: The southern grass skink lays approximately five eggs per clutch and is predominantly found in grassy treeless areas throughout its range.

Facts: Many reptiles that live in open grassy areas have body markings that include rows of alternating light and dark stripes. This type of body pattern is common in different reptile families and is most noticeable in species such as the canegrass dragon from central Australia and striped legless lizard from south-eastern Australia as well as many arid zone geckos in the genus *Strophurus.* The southern grass skink, as its name suggests, exhibits similar markings and is found in treeless areas in a narrow band extending from northern New South Wales to South

Australia. Striped body patterns break up the body form and provide camouflage from predators.

Management: Native grass, and leaf litter and fine woody debris.

Glossy grass skink, mountain bog skink

Pseudemoia rawlinsoni

David Hunter

Identifying features: Body colour brown to olive-green with a thin, black vertebral stripe. A dark brown lateral stripe is bordered above and below by two thin white stripes.

Snout–vent length: 62 mm.

Similar species: The woodland tussock skink has broken lateral stripes and the dorsolateral stripe on the southern grass skink is positioned on the fourth scale row from the vertebral stripe.

Life-form: Terrestrial.

Reproductive mode: Viviparous.

Activity: Diurnal.

Distribution: Restricted. Known only from Kosciuszko National Park.

Shelter-sites: Grass tussocks, mats of vegetation and fallen timber.

Vegetation type: Alpine woodland and montane forest.

Conservation status: Rare.

Additional notes: The glossy grass skink gives birth to live young and is predominantly found in high-elevation swamps and bogs. It is one of a suite of High Country species likely to be sensitive to changes in vegetation structure associated with climate change.

Facts: Species that live in thick vegetation such as swamps or wet heathland may find moving and escaping from predators a challenge. The glossy grass skink has smooth, shiny body scales compared to related members of the genus in the catchment. This trait enables them to glide through thick vegetation. This is in contrast to species that live in open rocky areas which usually have rough bodies and often raised scales. This helps them grip to rock surfaces and prevents predators from easily extracting them from crevices.

Management: Wetlands and native grass.

Kyalite
Moulamein
Conargo
Jerilderie
Urana
Swan Hill
Deniliquin
Finley
Holbrook
Tumbarumba
Barham
Tocumwal
Yarrawonga
Corowa
Albury
Khancoban
Moama

Scincidae

Spencer's skink, trunk-climbing skink

Pseudemoia spenceri

Identifying features: Body colour dark grey to black with pale flecks along the back and broad lateral stripe bordered above by row of small spots. The limbs and digits are elongated and body flat.

Snout–vent length: 65 mm.

Similar species: None.

Life-form: Arboreal and saxicolous.

Reproductive mode: Viviparous.

Activity: Diurnal.

Distribution: Localised. Predominantly found in the far eastern part of the Murray catchment, particularly Kosciuszko National Park.

Shelter-sites: Rock crevices and exfoliating bark.

Vegetation type: Alpine woodland, **montane forest** and **wet forest**.

Conservation status: Common.

Additional notes: Spencer's skink is a capable climber and is found on exposed rocky outcrops throughout the High Country.

Facts: All lizards need to raise their body temperature to complete other activities such as foraging, metabolising food and finding mates. Species that occur in areas with closed canopies may find this a challenge. Spencer's skink has overcome this problem by basking in gaps in the forest created by fallen trees or branches. It also occupies areas of mature trees. In these areas, canopy gaps and dappled sunlight is more plentiful than young, regenerating forest. They have even been recorded basking over 60 m above the ground in the upper branches of mountain ash, Australia's tallest tree.

Management: Rock outcrops, fallen timber and dead trees, and large mature trees.

Kyalite
Moulamein
Conargo
Jerilderie
Urana
Swan Hill
Deniliquin
Finley
Holbrook
Tumbarumba
Barham
Tocumwal
Yarrawonga
Corowa
Albury
Khancoban
Moama

Scincidae

Blotched blue-tongue, southern blue-tongue

Tiliqua nigrolutea

Damian Michael

Identifying features: Body colour black or dark brown with a series of pale cream to yellow blotches along the body. A blue tongue is displayed when threatened.

Snout–vent length: 300 mm.

Similar species: The eastern blue-tongue has bands along the body.

Life-form: Terrestrial.

Reproductive mode: Viviparous.

Activity: Diurnal.

Distribution: Localised. Found predominantly east of Jingellic in the Murray catchment but recently recorded in the north-western part of the Upper Billabong catchment east of Holbrook.

Shelter-sites: Fallen timber, debris and low shrubs.

Vegetation type: Montane forest and dry forest.

Conservation status: Common.

Additional notes: The blotched blue-tongue gives birth to between five and twelve young. It feeds on a range of invertebrates, including snails, as well as fungi and plant material.

Facts: How long a lizard survives in the wild depends on many factors including the prevalence of drought, prey abundance, habitat quality, proximity to roads, ability to avoid predators and aggressive encounters with other species. Generally, a lizards' life span is closely related to its body size. Small reptiles often live for just two to four years, while large species, particularly those in the *Tiliqua* genus, can live for 20 years, especially in captivity. It is for this reason that the blue-tongue lizard is commonly kept as a pet in Australia and overseas markets.

Management: Fallen timber and dead trees, and shrubs.

Damian Michael

Shingleback, stumpy-tail lizard

Tiliqua rugosa ssp. *aspera*

Identifying features: Body colour tan-brown, often with pale blotchy scales. The head is wide and triangular, body scales are raised and tail short and stumpy tail. A blue tongue is displayed when threatened.

Snout–vent length: 300 mm.

Similar species: None.

Life-form: Terrestrial.

Reproductive mode: Viviparous.

Activity: Diurnal.

Distribution: Localised. Predominantly found in the northern parts of the Murray catchment.

Shelter-sites: Beneath low growing shrubs and fallen timber.

Vegetation type: Sandhill woodland, **boree woodland**, **chenopod shrubland** and mallee woodland.

Conservation status: Uncommon, although locally abundant in the north of the region.

Additional notes: The shingleback gives birth to two or three large young and is a conspicuous part of the reptile fauna in arid parts of Australia. It feeds on a range of invertebrates, including snails, as well as fungi and plant material.

Facts: When it comes to reptile mating systems, most species are considered to be polygamous. That is, both males and females will mate with several partners over the breeding season. However, some species such as the shingleback form monogamous pairs and stay together between September and November before separating for the rest of the year. The same pair of lizards mates each year. But not all partners are 'faithful' as a small percentage of males will pair up with another

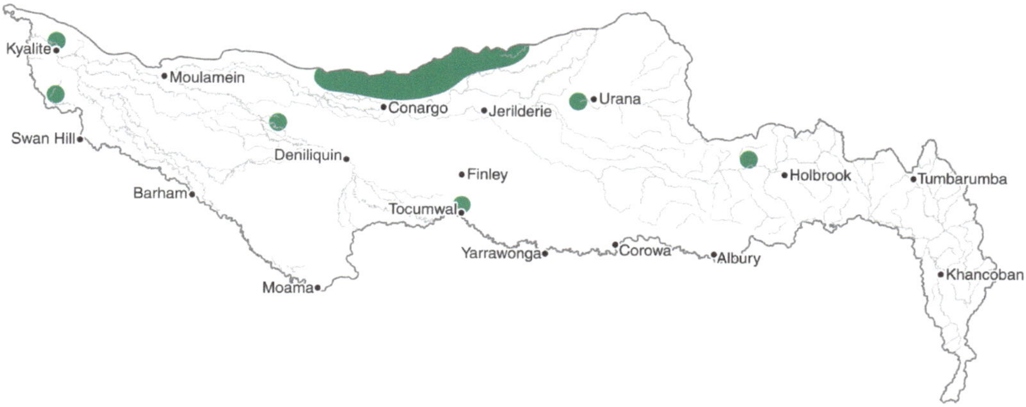

female. Females with promiscuous partners are themselves, more likely to mate with multiple males, giving rise to offspring with different fathers.

Management: Fallen timber and dead trees, and shrubs.

Damian Michael

Scincidae

Eastern blue-tongue, common blue-tongue

Tiliqua scincoides ssp. *scincoides*

Damian Michael

Identifying features: Body colour silvery-grey with a broad dark band from the eye to the ear and a series of dark cross bands along the body. A blue tongue is displayed when threatened.

Snout–vent length: 320 mm.

Similar species: The blotched blue-tongue has pale blotches along the body.

Life-form: Terrestrial.

Reproductive mode: Viviparous.

Activity: Diurnal.

Distribution: Widespread except the far eastern part of the Murray catchment.

Shelter-sites: Fallen timber and low growing shrubs.

Vegetation type: Dry forest, **grassy box woodland**, **sandhill woodland**, **black box woodland**, river red gum forest, **boree woodland** and chenopod shrubland.

Conservation status: Common.

Additional notes: The eastern blue-tongue is more commonly encountered in urban environments than bushland areas. In many towns and suburbs, the eastern blue-tongue will occupy overlapping home ranges that span several gardens, but each lizard usually confines its activities to a core area with a few favourite shelter-sites. Males usually have larger home ranges than females and pregnant females spend a lot more time in one location making them susceptible to being killed by pets.

Facts: Having a blue-tongue lizard in the garden can be useful in controlling snails and other pest animals (they have even

been known to eat baby mice). The species uses its powerful jaws to crack open the snail shell, which they often than discard using their fleshy tongue.

Management: Fallen timber and dead trees, shrubs, and exotic animals.

Blind snakes

Family Typhlopidae

The Family name Typhlopidae means 'blind eye' referring to the transparent scale that covers the eyes. However, they are technically not blind as they possess rudimentary vision and are able to detect light. They are found in central and southern America, south-eastern Europe, Asia, Australia and some western Pacific Islands. All blind snakes are fossorial and spend most of the time beneath leaf litter, inside rotting logs or below the soil surface. They are often encountered when soil or logs are disturbed or on the surface of the ground on warm summer nights, especially after rain. Australia has 42 species of blind snake which belong to a single genus *Ramphotyphlops*. The Latin name *Ramphotyphlops* means 'beaked blind eye' referring to the pointed snout in many species. Four species of blind snake are found in the Murray catchment.

Most blind snakes are difficult to identify and differ only in head shape and scale counts. However, the four species in the Murray catchment have broadly different habitat preferences and it is rare to encounter more than two species in the same area. Two species have broad distributions in the Murray catchment. The blackish blind snake is the largest species in the catchment and is found in moist environments east of the Hume Highway, whereas the prong-snouted blind snake is found widespread west of the Hume Highway. The woodland blind snake has a patchy distribution throughout the central part of the catchment and the southern blind snake is the most range-restricted of the four species and known only from the western part of the Murray catchment.

Typhlopidae

Dark-spined blind snake, southern blind snake

Ramphotyphlops bicolor (formerly *Ramphotyphlops australis*)

Matthew Herring

Identifying features: Body colour grey to purplish-brown with a rounded head which is angular when viewed from the side. Twenty-two mid-body scales.

Snout–vent length: 420 mm.

Similar species: The woodland blind snake has a weakly lobed-snout and 20 mid-body scales and the prong-snouted blind snake has a pronounced trilobed-snout and 20 mid-body scales.

Life-form: Fossorial.

Reproductive mode: Oviparous.

Activity: Nocturnal.

Distribution: Restricted. Known only from a few records north of Deniliquin.

Shelter-sites: Loose soil, leaf litter and ant nests.

Vegetation type: Sandhill woodland.

Conservation status: Rare.

Additional notes: The southern blind snake lays eggs and feeds principally on termites and ant larvae. The number of eggs it produces is unknown. The species may have a greater preference for sandier soils than other blind snakes, and may potentially occur in more areas in the western part of the Murray catchment.

Facts: Many snakes have modifications to their tails that can either help to deter predators or attract prey. For example, North American rattlesnakes use their tails to ward off potential predators. The Australian death adder has a modified soft spine at the tip of the tail that it uses to lure prey. Blind snakes also have modified tails which end in a short, sharp spine. This spine may serve two main functions. First, it may be used as an anchor point to help with movement within the tunnel systems of ants. Second, it may be used in defence when attacked by predators. When a blind snake is handled the spine is often jabbed into the hand.

Management: Soil structure and integrity, leaf litter and fine woody debris, and fallen timber and dead trees.

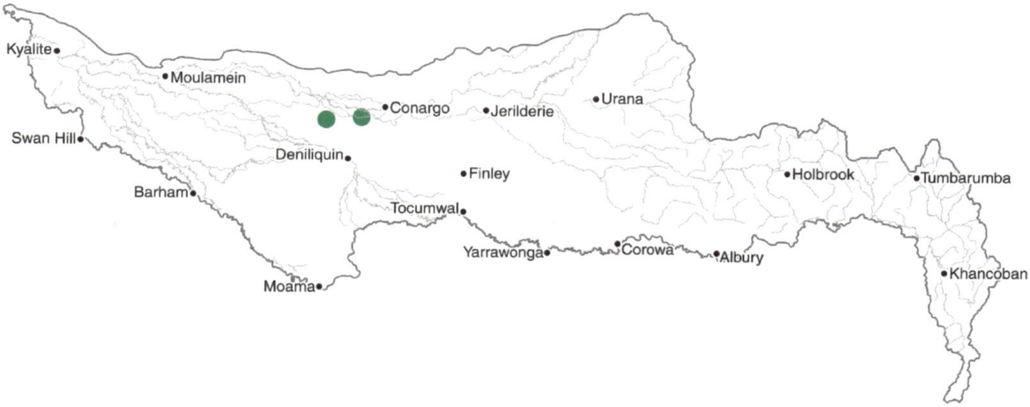
Kyalite
Moulamein
Conargo
Jerilderie
Urana
Swan Hill
Deniliquin
Finley
Holbrook
Tumbarumba
Barham
Tocumwal
Yarrawonga
Corowa
Albury
Khancoban
Moama

Typhlopidae

Prong-snouted blind snake, Peter's blind snake

Ramphotyphlops bituberculatus

David Webb

Identifying features: Body colour pink to brown and the head is angular in profile with three distinctive lobes on the snout. Twenty mid-body scales.

Snout–vent length: 450 mm.

Similar species: The woodland blind snake has a weakly lobed-snout and 20 mid-body scales and the southern blind snake has 22 mid-body scales and lacks a lobed-snout.

Life-form: Fossorial.

Reproductive mode: Oviparous.

Activity: Nocturnal.

Distribution: Widespread except for the eastern part of the Murray catchment.

Shelter-sites: Fallen timber, fence posts and ant nests.

Vegetation type: Grassy box woodland, **sandhill woodland**, **boree woodland** and mallee woodland.

Conservation status: Common.

Additional notes: The prong-snouted blind snake lays eggs and feeds principally on termites and ant larvae. The number of eggs it produces is unknown. The species is found on a range of soils types ranging from sand to hard clay. In areas with heavy soils it is likely to rely on soil cracks and ant nests for shelter.

Facts: The shape of the head when viewed from the side and above is one of the main ways of differentiating between the 42 different species of blind snake in Australia. Some species have round snouts, whereas others, such as the prong-snouted blind snake, have angular profiles and noticeably lobed-snouts. In general, species that live in cool, moist parts of eastern Australia have typically rounded-snouts, whereas species that inhabit arid areas have lobed-snouts. It is likely that having a lobed-snout helps them push through heavy soils and may even serve to protect their nostrils from filling up with soil.

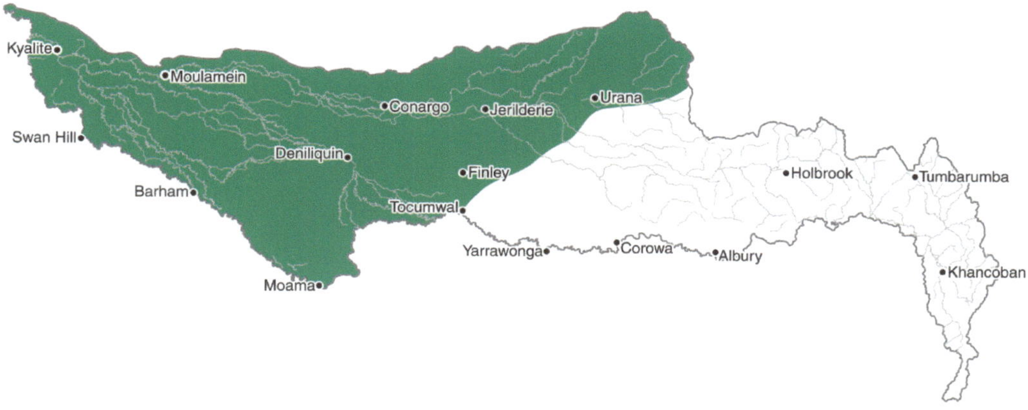

Management: Soil structure and integrity, leaf litter and fine woody debris, and fallen timber and dead trees.

Peter Robertson

Blackish blind snake

Ramphotyphlops nigrescens

Identifying features: Body colour pinkish-grey to almost black and the head is rounded from above and the side. Twenty-two mid-body scales.

Snout–vent length: 750 mm.

Similar species: The woodland blind snake has a weakly lobed-snout and 20 mid-body scales and the prong-snouted blind snake has distinctive lobed-snout and 20 mid-body scales.

Life-form: Fossorial.

Reproductive mode: Oviparous.

Activity: Nocturnal.

Distribution: Localised. Predominantly found east of Woomargama National Park.

Shelter-sites: Leaf litter, fence posts, embedded rocks and ant nests.

Vegetation type: Wet forest, dry forest and grassy box woodland.

Conservation status: Uncommon.

Additional notes: The blackish blind snake produces up to 19 eggs per clutch and feeds principally on the larvae of large ant species. Aggregations of up to a dozen individuals have been recorded.

Facts: All blind snakes are non-venomous. They are completely harmless to humans and have never been known to even attempt to bite. Instead, when handled or attacked by a predator they often emit a foul-smelling odour from glands located near the base of their tail. Large species, such as the blackish blind snake can exude large amounts of this substance and in quantities that prevent prolonged human handling.

Management: Soil structure and integrity, leaf litter and fine woody debris, and fallen timber and dead trees.

Peter Robertson

Typhlopidae

Woodland blind snake

Ramphotyphlops proximus

Damian Michael

Identifying features: Body colour pink to dark greyish-brown and the head is slightly angular and weakly trilobed. Twenty mid-body scales.

Snout–vent length: 750 mm.

Similar species: The prong-snouted blind snake has pronounced trilobed-snout and 20 mid-body scales. The blackish blind snake has 22 mid-body scales and a rounded snout.

Life-form: Fossorial.

Reproductive mode: Oviparous.

Activity: Nocturnal.

Distribution: Widespread but patchily distributed across the Murray catchment. The species is absent from the far west of the region.

Shelter-sites: Embedded rocks, fallen timber and ant nests.

Vegetation type: Grassy box woodland and sandhill woodland.

Conservation status: Rare. The species is listed as 'vulnerable to extinction' in Victoria.

Additional notes: The woodland blind snake produces eggs and feeds principally on termites and ant larvae. The number of eggs it produces is unknown. In the eastern part of its range it is often encountered beneath deeply embedded granite rocks and may prefer environments with coarse-grained soil.

Facts: Geckos are well known for their vocal abilities and many species, particularly the introduced Asian house gecko, are often heard making 'chuck chuck' noises during at night in northern Australia. Many species of legless lizard are also capable of making audible sounds, but the only sound snakes are capable of making are 'hissing'

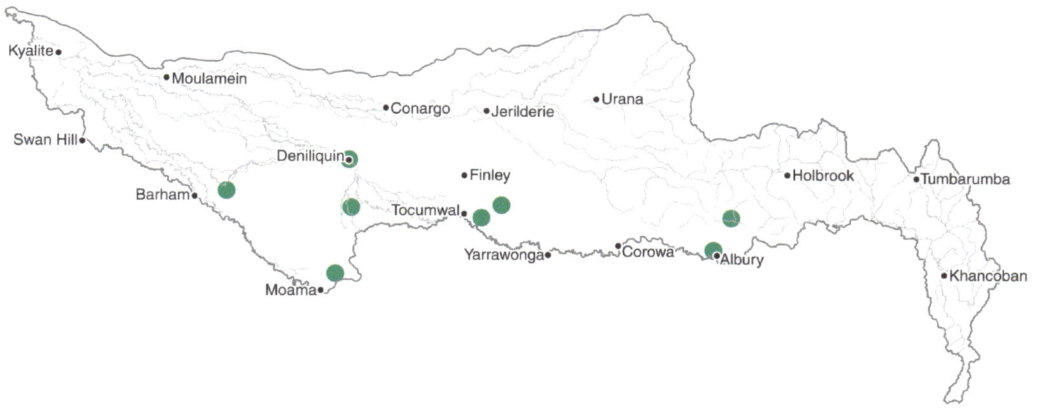

noises made by expelling air from the lungs. Blind snakes are an exception and species such as the woodland blind snake are capable of emitting an audible squeak when handled. The reason for their vocal abilities are unknown as these species spend most of their lives beneath the ground and are not thought to be territorial or show any aggressive behaviour towards other species.

Management: Soil structure and integrity, leaf litter and fine woody debris, and fallen timber and dead trees.

Damian Michael

Pythons

Family Pythonidae

Pythons are found in Africa, India, South-East Asia and Australia and reach their greatest diversity in Australasia, particularly throughout the Australo–Papua region. They are distinguished from other snakes in Australia by their large body sizes and heat sensitive pits (in most species) on the upper and lower lips. Pythons are among the longest snakes in the world and include such giants such as: the African rock python, Asian rock python, reticulated python and the Australian scrub python. All of these species can attain 6 metres in length and the reticulated python may exceed 11 metres. Australia has 13 species of python which belong in four genera. The most diverse genus is *Morelia* which contains five species and one species, *Morelia spilota* contains six subspecies.

One species of python, *Morelia spilota* ssp. *metcalfei* occurs in the Murray catchment. Anecdotal reports by landholders in the SWS bioregion claim pythons were introduced to hay and grain sheds following the Second World War. These introductions were to control rats and may have included *Morelia spilota* ssp. *metcalfei* from northern parts of the Murray–Darling catchment and southern Queensland, or the coastal subspecies, *Morelia spilota* ssp. *mcdowelli*. However, it remains to be established whether coastal forms still occur in rural areas in the Murray catchment. Furthermore, the coastal carpet python and another subspecies, the diamond python *Morelia spilota* ssp. *spilota*, are commonly kept in captivity as pets. Occasionally, these snakes escape from their enclosures and are infrequently encountered in urban areas, including Albury.

Inland carpet python, Murray-Darling carpet python

Morelia spilota ssp. *metcalfei*

Mason Crane

Identifying features: Body colour pale grey with reddish-brown tinge to the sides. A series of irregularly paired, pale blotches along the back and pale stripes along each side. Lip scales have heat sensitive pits.

Total length: 2.5 m.

Similar species: None.

Life-form: Arboreal, saxicolous and terrestrial.

Reproductive mode: Oviparous.

Activity: Nocturnal but often seen basking in the early morning and late afternoon.

Distribution: Widespread but patchily distributed in the SWS bioregion and floodplain forests in the western part of the Murray catchment. Known from a record as far east as Ournie in the upper Murray region.

Shelter-sites: Tree hollows, hollow logs, rock crevices, rabbit burrows and rural buildings.

Vegetation type: Dry forest, **grassy box woodland**, sandhill woodland, **black box woodland**, **red gum forest** and mallee woodland.

Conservation status: Uncommon. Populations have severely declined across the Murray catchment.

Additional notes: The inland carpet python produces approximately 20 eggs per clutch and incubates them for an average of 70 days. It reproduces every three years and feeds on a range of vertebrates including, mice, rats, rabbits, possums and nestling birds. Juvenile pythons prey on small lizards.

Facts: The apparent decline of this species may in part be due to declining rabbit

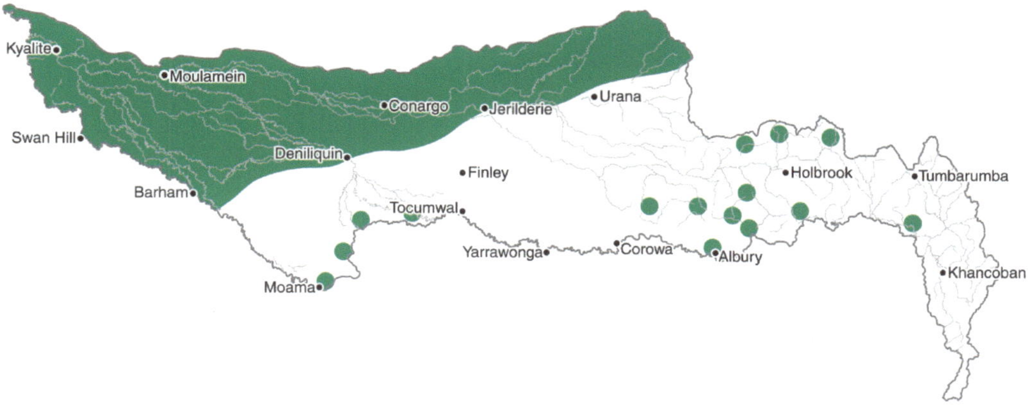

numbers as well as habitat loss. In many areas, the inland carpet python feeds entirely on rabbits. In the SWS bioregion, relict populations are restricted to large, granite outcrops where they find suitable shelter and food. In the west, they are restricted to river red gum forests and well-vegetated creeks. There are few records of juvenile snakes in the catchment and it is likely that populations once inhabiting valleys in the upper part of the catchment are now locally extinct.

Management: Rocky outcrops, leaf litter and fine woody debris, fallen timber and dead tree, shrubs, large mature trees, invasive exotic plants, and exotic animals.

Damian Michael

Venomous land snakes

Family Elapidae

Elapid (front-fanged) snakes are found in the Americas, Africa, the Middle East, Asia and Australia and reach their greatest diversity in Australasia, where they are thought to have originated. They are distinguished from other land-based snakes by small fangs, positioned near the front of the mouth. Snakes in this family include the world's most venomous reptile species, including Asian cobras *Naja* spp., African mambas *Dendroaspis* spp. and numerous Australian snake species. Australia has 97 species of venomous land snakes which belong in 26 different genera. The most species-rich groups include, the shovel-nosed snakes *Brachyurophis* spp., whipsnakes *Demansia* spp., hooded snakes *Parasuta* spp., black snakes *Pseudechis* spp. and brown snakes *Pseudonaja* spp. Together they comprise over 40% of the venomous snake fauna in Australia. Thirteen of these species occur in the Murray catchment.

Snakes can vary considerably in body colour. However, with the exception of a few species, identifying snakes in the Murray catchment should not be difficult. Six species, the death adder, eastern small-eyed snake, yellow-faced whip snake, red-naped snake, western brown snake and bandy bandy are known from less than five records. In contrast, some species such as the eastern brown snake and red-bellied black snake are widespread and regularly encountered. Others, such as the highlands copperhead and white-lipped snake are found only in the far eastern part of the Murray catchment. With the exception of the bandy bandy, all elapid snakes in the Murray catchment should be treated with caution and handled only by experienced herpetologists. Fatalities arising from snake bites are rare in Australia. However, the chance of being bitten by a venomous species increases when people attempt to kill them. Sadly, many snake species in the catchment that have never caused a fatality are deliberately killed because they are mistaken for an eastern brown snake. If a snake is encountered, the best thing to do is to retreat slowly and avoid making rapid movements. Generally, most snakes will retreat to cover or give a warning sign that they feel threatened. This often involves flattening of the neck, raising the head off the ground and hissing.

Common death adder

Acanthophis antarticus

Peter Robertson

Identifying features: Body colour grey to reddish-brown with lighter cross bands along the body. Head is broadly diamond-shaped, body short and stocky and is the only species with a tail ending in a soft spine.

Total length: 700 mm.

Similar species: The eastern tiger snake and eastern blue-tongue may have similar body markings. However, both of these species lack a soft spine on the tail.

Life-form: Terrestrial.

Reproductive mode: Viviparous.

Activity: Nocturnal but occasionally encountered moving during the day, especially prior to summer storms.

Distribution: Restricted. Known only from two records near Moulamein. In 2007, a landholder reported seeing a specimen crossing a road near Strathmerton in Victoria.

Shelter-sites: Leaf litter and fallen timber.

Vegetation type: Sandhill woodland.

Conservation status: Rare.

Additional notes: The death adder gives birth to between two and 24 young. It feeds on a range of vertebrates including lizards, mammals and birds. Anecdotal records suggest that, historically, the death adder was more common and had a wider distribution, particularly in the northern part of the Murray catchment. It is extremely sensitive to disturbances such as livestock grazing and altered fire regimes and is likely to have disappeared from many parts of the catchment. The death adder has extremely long fangs (6 mm), second only to taipans. It is highly venomous and has been involved in several fatalities.

Facts: The death adder is a cryptic species that spends much of the time concealed beneath leaf litter using its modified tail as a lure to attract prey. The name death adder

is thought to be a corruption of the phrase 'deaf adder', alluding to its sit and wait hunting technique and reluctance to move unless provoked.

Management: Leaf litter and fine woody debris, surface rocks, and shrubs.

Elapidae

Highlands copperhead

Austrelaps ramsayi

Damian Michael

Identifying features: Body colour dark brown to grey with white triangular-shaped markings on upper lip scales.

Total length: 1.1 m.

Similar species: The white-lipped snake has a continuous white stripe on the upper lip scales.

Life-form: Terrestrial.

Reproductive mode: Viviparous.

Activity: Diurnal.

Distribution: Localised. Found only in the far eastern part of the Murray catchment, often above 700 metres elevation.

Shelter-sites: Fallen timber, hollow logs and debris.

Vegetation type: Alpine grassland, **alpine woodland**, **montane forest**, wet forest and dry forest.

Conservation status: Common.

Additional notes: The highlands copperhead gives birth to an average of 20 live young from mid-summer to early autumn. It can be common in alpine meadows and bogs and is often encountered basking in small aggregations. The highlands copperhead is highly venomous, although reluctant to bite. It has been known to catch and eat grasshoppers.

Facts: During the breeding season it is not uncommon to encounter a pair of male copperhead snakes entwined in a one-on-one ritual territorial battle to access receptive females. These combats are unusual in that the two snakes usually do not bite each other. Instead, their heads are kept well separated and only the lower halves of their bodies are entwined. The dominant, often heavier, individual tries to pin his opponent to the ground. Battles between

closely matched individuals can often last for as long as half an hour, ending only when one snake gives in.

Management: Native grass, wetlands, rocky outcrops, and fallen timber and dead trees.

David Hunter

Elapidae

Eastern small-eyed snake

Cryptophis nigrescens

Damian Michael

Identifying features: Body colour glossy black or dark grey with pinkish belly that is not visible from the side. It has small eyes and a rounded head.

Total length: 500 mm.

Similar species: The red-bellied black snake has red scales that extend higher up the body and are visible from the side.

Life-form: Terrestrial and saxicolous.

Reproductive mode: Viviparous.

Activity: Nocturnal.

Distribution: Restricted. Known only from a few records in Woomargama National Park.

Shelter-sites: Fallen timber, exfoliating rocks and surface rocks.

Vegetation type: Dry forest.

Conservation status: Rare.

Additional notes: The small-eyed snake is secretive and nocturnal and gives birth to eight live young. It feeds almost exclusively on skinks which are captured at night when they are sleeping. There are few records of this species in the Murray catchment, although it is likely to be more widespread in the upper part of the catchment, especially in areas with exfoliating granite slabs, which provide thermally suitable shelter-sites. It is venomous and has been responsible for causing a fatality in Victoria.

Facts: The small-eyed snake is extremely sensitive to light and will immediately try to hide its head when disturbed. It is possible for this species to be mistaken for the similar looking red-bellied black snake, despite differences in activity periods and preferred habitat. The small-eyed snake can be found in small aggregations and in some areas up to a dozen individuals have been found sequestered together beneath rock slabs and even behind the walls of old houses.

Management: Rocky outcrops, surface rocks, and exotic animals.

Peter Robertson

Elapidae

Yellow-faced whipsnake

Demansia psammophis ssp. *psammophis*

Peter Robertson

Identifying features: Body colour pale grey to olive-brown often with faint reddish lateral stripe. It has a black 'comma'-shaped marking covering each eye and a black banded-snout.

Total length: 1 m.

Similar species: None.

Life-form: Terrestrial.

Reproductive mode: Oviparous.

Activity: Diurnal.

Distribution: Restricted. Known only from Nail Can Hill Flora and Fauna Reserve in Albury, Benambra National Park, Woomargama National Park, Tumbarumba and near Tocumwal. In 2008, a specimen was recorded near Mount Mittamatite in Victoria, presumably the first record for north-eastern Victoria.

Shelter-sites: Fallen timber and surface rocks.

Vegetation type: Dry forest, grassy box woodland, sandhill woodland.

Conservation status: Rare, although potentially common in localised areas.

Additional notes: The yellow-faced whipsnake produces up to nine eggs per clutch often in communal nesting-sites and is often found basking in small aggregations. It is a venomous species, although only large individuals are likely to produce bite symptoms in humans.

Facts: Vision in snakes can be variable and is often dependant on where they live (i.e. below or above ground) and what time of the day they are active. The yellow-faced whipsnake has one of the largest eyes in relation to other snakes in the elapid family. Their well-developed eyesight enables them to forage for fast moving diurnal lizards such as *Ctenotus* species. Its agility and keen eyesight make it a difficult species to encounter in the wild as it will usually detect a potential threat well before it is seen. It may be for this reason that it is rarely detected in the catchment. It was not until 2007 that the first specimens were recorded in bushland near Albury.

Management: Native grass, surface rocks, and fallen timber and dead trees.

Damian Michael

White-lipped snake

Drysdalia coronoides

David Hunter

Identifying features: Body colour grey to dark brown with a black-edged white stripe running along the upper lip scales and onto the neck.

Total length: 450 mm.

Similar species: The highlands copperhead has broken white markings confined to the upper lip scales.

Life-form: Terrestrial.

Reproductive mode: Viviparous.

Activity: Diurnal but nocturnally active on warm nights.

Distribution: Localised. Predominantly restricted to Kosciuszko National Park.

Shelter-sites: Grass tussocks and fallen timber.

Vegetation type: **Alpine grassland**, **alpine woodland**, **montane forest**, wet forest and dry forest.

Conservation status: Common.

Additional notes: The white-lipped snake gives birth to a maximum of eight young during spring. It is often found in small aggregations basking on grass tussocks. It feeds on small lizards and frogs and although venomous, not considered dangerous. However, it can be pugnacious and has a tendency to bite if roughly handled.

Facts: Aggregations of snakes basking have been recorded in a variety of species. The reasons certain species bask communally may involve access to thermally suitable basking sites in areas that are cold, access to abundant prey, or safety in numbers ('more eyes the better'). The white-lipped snake is one of Australia's most cold-adapted reptile species. It can be active during the coolest days in spring and is one only three species of snake found in Tasmania. It is not unusual to encounter large

numbers of white-lipped snakes basking on the edges of swamps in the High Country, particularly at the beginning of spring and at the end of autumn. During the winter, however, the white-lipped snake will retreat underground, or communally over-winter beneath fallen logs.

Management: Native grass, wetlands, and fallen timber and dead trees.

Red-naped snake
Furina diadema

Peter Robertson

Identifying features: Body colour red-brown with a shiny, black head and nape completely enclosing a red or orange patch.

Total length: 400 mm.

Similar species: Juvenile eastern brown snakes have a neck band that is separated from the dark hood by a lighter band (some individuals have multiple dark bands along the body).

Life-form: Terrestrial and fossorial.

Reproductive mode: Oviparous.

Activity: Nocturnal.

Distribution: Restricted. Known only from near Woomargama National Park and Mullengandra.

Shelter-sites: Surface rocks, fallen timber, ants and termite nests.

Vegetation type: Dry forest and grassy box woodland.

Conservation status: Rare.

Additional notes: The red-naped snake produces up to five eggs per clutch and preys principally on small lizards. It is often encountered in ant and termite nests, and rotting timber. It is venomous but not considered dangerous and is reluctant to bite. It has also been recorded near Tumut and may be more common in dry forests in the eastern part of the Murray catchment.

Facts: The red-naped snake can be easily mistaken for a juvenile eastern brown snake. However, unlike the eastern brown snake they are rarely encountered during the day. Its burrowing habits make it difficult to detect, and like other fossorial species, it is often only encountered when disturbed from its shelter-site. The red-naped snake will often lift the front third of its body off the ground and make mock striking movements with its mouth closed.

This behaviour makes this inoffensive species appear threatening and may end in its untimely death.

Management: Soil structure and integrity, leaf litter and fine woody debris, surface rocks, and fallen timber and dead trees.

Elapidae

Eastern tiger snake, common tiger snake

Notechis scutatus

Jeremy Tscharke

Identifying features: Body colour olive, grey-brown or black with pale bands along the body.

Total length: 2 m.

Similar species: The bandy bandy has black and white bands and some juvenile eastern brown snakes have dark hoods and numerous narrow, dark cross bands.

Life-form: Terrestrial but occasionally found in low shrubs.

Reproductive mode: Viviparous.

Activity: Diurnal but can be active on warm nights.

Distribution: Widespread but patchily distributed in the western part of the Murray catchment.

Shelter-sites: Fallen timber and cavities below ground.

Vegetation type: Alpine woodland, **montane forest**, **wet forest**, dry forest, grassy box woodland, **river red gum forest** and **wetlands**.

Conservation status: Uncommon. Populations are reported to have declined in the Riverina bioregion.

Additional notes: The eastern tiger snake produces up to 40 live young. It feeds predominantly on frogs and lizards, although birds and mammals are also preyed upon. It is a highly venomous species and has been involved in several fatalities.

Facts: The eastern tiger snake has had a chequered history in the region. It was once abundant along the Murray River floodplain and adjacent wetlands throughout the Riverina. Persecution by 'sportsman' in the early 1900s and changes to flooding regimes have significantly

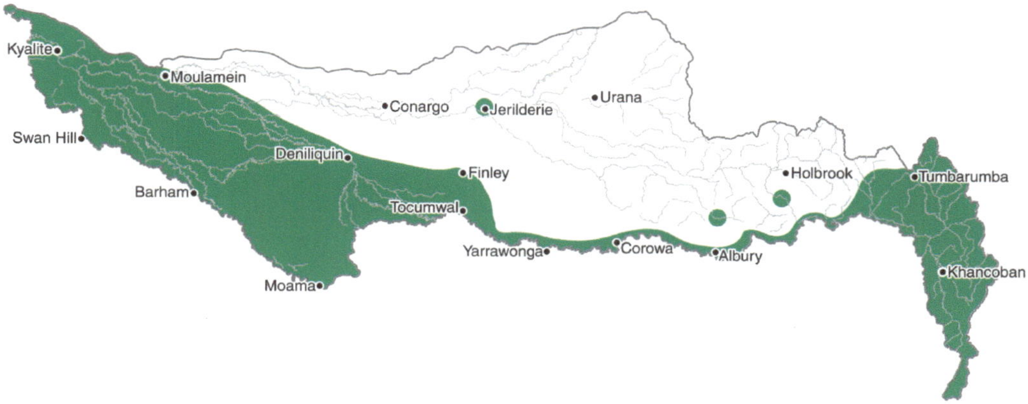

reduced its range and abundance. When irrigation channels were constructed and rice farming began, the eastern tiger snake adapted to this new environment, and in some areas numbers have recovered. It appears the long-term survival of the eastern tiger snake in large areas of the western part of the catchment may now be dependent on irrigation or environmental water to provide suitable habitat for this species (and its prey).

Management: Wetlands, and fallen timber and dead trees.

Dwyer's snake
Parasuta dwyeri

Identifying features: Body colour tan-brown with a dark hood covering both the head and neck.

Total length: 600 mm.

Similar species: Juvenile eastern brown snakes have a neck band that is separate to the dark hood (and occasionally multiple bands along the body). The curl snake has ragged, white-edged upper lip scales and often a light coloured scale between the eye and nose.

Life-form: Terrestrial.

Reproductive mode: Viviparous.

Activity: Nocturnal.

Distribution: Localised. Predominantly restricted to the SWS bioregion.

Shelter-sites: Fallen timber, surface rocks and exfoliating rock slabs.

Vegetation type: Dry forest, **grassy box woodland** and sandhill woodland.

Conservation status: Uncommon.

Additional notes: Dwyer's snake gives birth to five young and mating may occur in spring and autumn. It will quickly retreat down tunnels or cracks in the ground beneath a shelter-site when disturbed. It is venomous, although not considered dangerous.

Facts: Many reptiles in temperate regions are dormant for most days during the winter months. Dwyer's snake is one of the few snake species that continues to be active during the winter months. During the day, they raise their body temperature by pressing themselves into tight fitting cracks in dead timber or rock exfoliations. Being in close contact with a warm substrate situated in full sun enables Dwyer's snake to

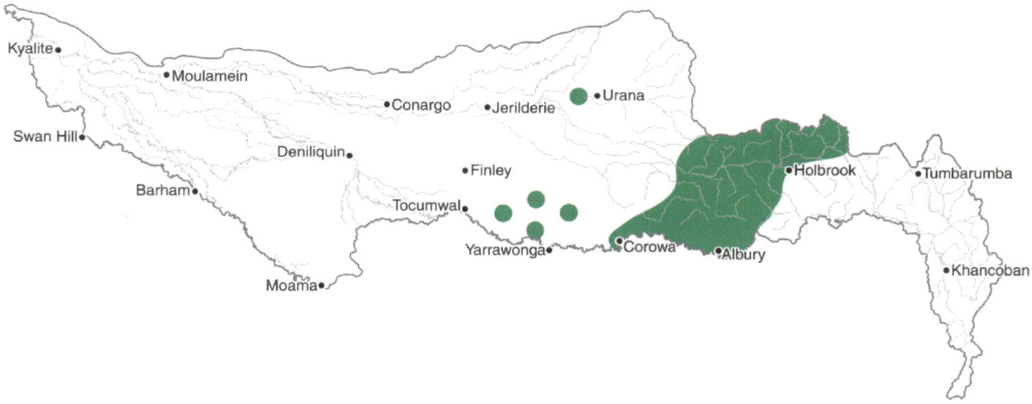

heat itself sufficiently to then forage into the night and prey on sleeping lizards. Superficial resemblance to juveniles of the eastern brown snake may result in many animals being misidentified and killed.

Management: Rocky outcrops, surface rocks, and fallen timber and dead trees.

Damian Michael

Elapidae

Red-bellied black snake

Pseudechis porphyriacus

David Blair

Identifying features: Body colour black and shiny with a reddish belly.

Total length: 2 m.

Similar species: The eastern small-eyed snake has pinkish-red belly scales not extending up the sides of the body.

Life-form: Terrestrial and semi-aquatic.

Reproductive mode: Viviparous.

Activity: Diurnal.

Distribution: Localised. Predominantly found along the Murray, Wakool and Edward river systems but also along creek systems in upper Murray region.

Shelter-sites: Fallen timber, cavities below ground and large rock slabs.

Vegetation type: Dry forest, grassy box woodland, **river red gum forest** and **wetlands**.

Conservation status: Common.

Additional notes: The red-bellied black snake gives birth to an average of 20 young which emerge from membranous sacs within an hour of being delivered. The species is often associated with wet environments and has been observed foraging beneath water in search of tadpoles, frogs or fish. It is venomous and, although no human fatalities have been recorded, bites to the hand often result in amputation of the fingers due to the substantial tissue damage that is caused by compounds in the venom.

Facts: The red-bellied black snake is typically a placid and inoffensive animal that is reluctant to bite unless severely provoked. However, if provoked it will flatten its neck and turn side on to the attacker while retreating to cover. It is also capable of delivering a bite without injecting venom. These 'dry bites' are usually delivered when they are accidentally stepped on. In other cases, the red-bellied black snake will recoil their body hitting a

person's legs with the tail, giving the impression of being bitten.

Management: Wetlands, and fallen timber and dead trees.

Elapidae

Western brown snake

Pseudonaja nuchalis

Peter Robertson

Identifying features: Body colour olive-grey to brown with or without dark cross bands along the body and dark markings on the neck. The snout scale is wide and 'strap-like' and mouth lining is purplish-black.

Total length: 1.6 m.

Similar species: The eastern brown snake has a pink mouth lining and typically no adult body markings.

Life-form: Terrestrial.

Reproductive mode: Oviparous.

Activity: Diurnal although active on warm nights.

Distribution: Restricted. Known only from near Deniliquin in the central part of the region.

Shelter-sites: Fallen timber, shrubs and ground cavities.

Vegetation type: Sandhill woodland and boree woodland.

Conservation status: Rare.

Additional notes: The western brown snake produces up to 30 eggs per clutch and feeds on small lizards and mammals. It is highly venomous and has a nervous temperament. If provoked, it will raise the fore body off the ground to form an 'S' shape and open its mouth. A bite from this species may cause severe symptoms resulting in death.

Facts: Body colour can often be a poor way to identify snakes. Many species vary considerably in body colour and markings within single populations and across their geographical ranges. The western brown snake is a classic example of just how variable snakes can be. Dozens of colour forms and body patterns have been described. Some may represent undescribed subspecies while others may even be undescribed species. The only way to accurately identify particular snake species is to count the scales around the mid-body and along the lips. This should only be attempted by

experienced herpetologists, as a scratch from the fangs of a dead snake can still contain enough residual venom to be fatal.

Management: Native grass, and fallen timber and dead trees.

Eastern brown snake

Pseudonaja textilis

David Webb

Identifying features: Body colour tan-brown to dark brown usually without body marking but occasionally faintly banded adults occur. Juveniles have a black head and neck band and occasionally multiple dark bands along the body.

Total length: 2.2 m.

Similar species: The western brown snake has a purplish mouth lining and dark neck markings.

Life-form: Terrestrial.

Reproductive mode: Oviparous.

Activity: Diurnal.

Distribution: Widespread except for the far eastern part of the Murray catchment.

Shelter-sites: Fallen timber, ground cavities and surface rocks.

Vegetation type: Dry forest, **grassy box woodland**, **sandhill woodland**, river red gum forest, **boree woodland**, **chenopod shrubland** and mallee woodland.

Conservation status: Common. This species may have benefited from human activities.

Additional notes: The eastern brown snake produces up to 30 eggs per clutch. It feeds on lizards and mammals. It is highly venomous and has a nervous temperament. If provoked, it will raise the fore body off the ground to form an 'S' shape and open its mouth. A bite from this species may cause severe symptoms. This species is responsible for the majority of snake bite-related deaths in Australia.

Facts: Elapid snakes are thought to have arrived in Australia during the Pliocene

epoch approximately 12–14 million years ago. At this time, sea levels were much lower than today and land bridges linked Australia via Papua New Guinea to Asia. It was during this period that a cobra-like ancestor entered Australia and diversified into the approximately 100 different front-fanged snakes that exist today. This process is called adaptive radiation.

Management: Native grass, rocky outcrops, surface rocks, and fallen timber and dead trees.

Peter Robertson

Elapidae

Curl snake, Myall snake

Suta suta

Damian Michael

Identifying features: Body colour light to dark brown with a white lip bordered above by a dark ragged-edged stripe and dark brown or black hood.

Total length: 600 mm.

Similar species: Juvenile eastern brown snakes have a separate dark neck band and Dwyer's snake lacks white upper lip scales.

Life-form: Terrestrial.

Reproductive mode: Viviparous.

Activity: Nocturnal.

Distribution: Widespread in the western part of the Murray catchment. Found as far east as Urana.

Shelter-sites: Fallen timber and soil cracks.

Vegetation type: Sandhill woodland, **black box woodland**, river red gum forest, **boree woodland** and **chenopod shrubland**.

Conservation status: Common. This species is listed as 'vulnerable to extinction' in Victoria.

Additional notes: The curl snake gives birth to seven young and is common in areas with cracking clay soils where it feeds on geckos and small lizards. This species will quickly colonise artificial habitats such as old fence posts and railway sleepers and can often be found by spotlighting during warm nights. It is a venomous species and a bite from large individuals may cause severe symptoms such as swelling and blistering of the lymph nodes.

Facts: The curl snake gets its name from its tendency to rapidly curl up its body and whip from side to side. This kind of behaviour is shared with a number of other small, brown, hooded snakes in south-eastern Australia. If disturbed, the curl snake will escape into a hole or crack in the ground.

Management: Soil structure and integrity, native grass, leaf litter and fine woody debris, fallen timber and dead trees, and exotic animals.

Damian Michael

Elapidae

Eastern bandy bandy

Vermicella annulata

Jerry Alexander

Identifying features: Body colour distinctively banded black and white.

Total length: 760 mm.

Similar species: None, although, juvenile eastern brown snakes occasionally have thin, dark cross bands along the body.

Life-form: Fossorial.

Reproductive mode: Oviparous.

Activity: Nocturnal.

Distribution: Restricted. Known only from Deniliquin, Wakool, Tocumwal and Goombargana Hill near Walbundrie.

Shelter-sites: Leaf litter, loose soil, rock crevices, surface rocks and fallen timber.

Vegetation type: Grassy box woodland and sandhill woodland.

Conservation status: Rare. Populations have declined in the Murray catchment. The species is listed as 'near threatened' in Victoria.

Additional notes: The bandy bandy produces up to 13 eggs. It is only active above the ground on warm summer nights, particularly after heavy rain. The species feeds principally on blind snakes and is capable of detecting and following their chemical scent trails. The bandy bandy is venomous but not considered dangerous and makes no attempt to bite.

Facts: When threatened, the bandy bandy performs an unusual behaviour of lifting sections off its body of the ground to form one or two loops. This behaviour may have given rise to the myth that when threatened, snakes hold onto their tails with their

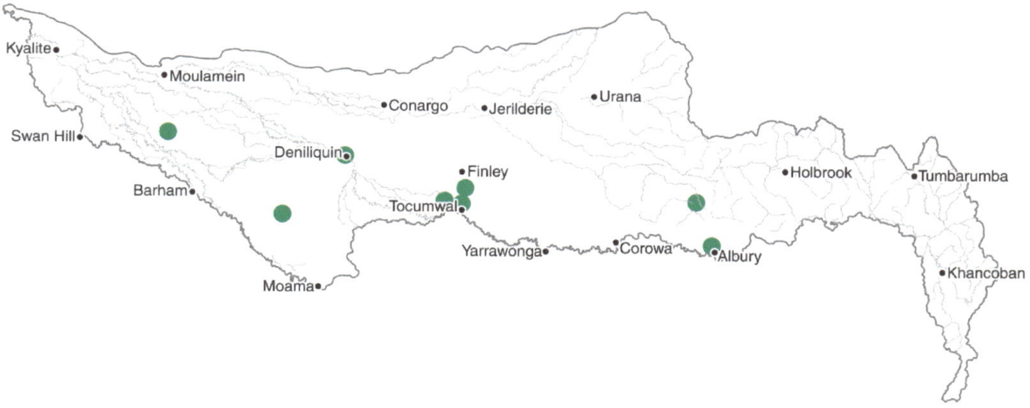

mouth, form a hoop and roll away. The bandy bandy also uses 'flicker fusion' to confuse its predators. When it moves about at night, the black and white body markings appear to blur together, making the body outline difficult to discern, and the snake difficult to catch.

Management: Soil structure and integrity, rocky outcrops, leaf litter and fine woody debris, surface rocks, fallen timber and dead trees, and exotic animals.

Jeremy Tscharke

6 Identifying similar looking skinks in the Murray catchment

Many small lizards look alike and can be extremely difficult to tell apart in the field. Because of this we provide a summary of the key distinguishing features that will help readers identify similar looking skink species in the Murray catchment. Below, we outline the main identifying features of individual species that are unique among the following groups: 1) fossorial species, including *Anepischtos*, *Hemiergis* and *Lerista*; 2) fasting-moving *Ctenotus* species; 3) rock-dwelling *Egernia* species; 4) rock-associated *Liopholis* species; 4) semi-aquatic *Eulamprus* species; and 5) High Country *Pseudemoia* species. Appendix 1 provides a list of all reptile species and the main vegetation types they are associated with in the Murray catchment.

Fossorial species (elongated and limb-reduced, ground-dwelling skinks)

There are five fossorial species in the Murray catchment. These species are distinguished by their digit arrangement and belly colour. Two species, McCoy's skink and the three-toed skink, are found east of the Hume Highway and co-occur in the High Country. In the SWS bioregion, Bougainville's skink overlaps with the three-toed skink east of the Hume Highway and with the wood-mulch slider west of the Hume Highway. The eastern robust slider and wood-mulch slider co-occur in mallee woodland in the far western part of the catchment. All species burrow below ground when disturbed.

Ctenotus species (fast-moving 'striped skinks')

There are six *Ctenotus* species in the Murray catchment. These species are distinguished by the combination of spots and stripes along the body. The Murray striped skink and royal skink co-occur in mallee woodland in the far western part of the catchment. The large striped skink co-occurs with the copper-tailed skink in the SWS bioregion and potentially with the spotted-back skink and brown-blazed

Table 1 Key distinguishing features between five fossorial skink species in the NSW Murray catchment.

Species		Digit arrangement	Belly colour	Page no.
McCoy's skink		Five fingers, five toes	Yellow	100
Three-toed skink		Three fingers, three toes	Yellow	132
Bougainville's skink		Five fingers, five toes	White - pale grey	138
Eastern robust slider		One finger, two toes	White - pale yellow	140
Wood-mulch slider		Three fingers, three toes	White - pale grey	142

wedgesnout skink in the Riverina. These skinks are among the fasted moving reptiles in Australia and will flee to nearby cover when disturbed.

Egernia species (large, rough-bodied, rock-dwelling skinks)

There are three *Egernia* species in the Murray catchment. These species are distinguished by their body texture and side markings. Cunningham's skink co-occurs with the other two *Egernia* species. The tree crevice skink and black rock skink are not expected to co-occur in the Murray catchment. All three species will retreat within rock crevices when disturbed.

Liopholis species (medium-sized, smooth-bodied skinks)

There are three *Liopholis* species in the Murray catchment. These species can be distinguished by the combination of dorsal and lateral markings. The tan-backed

Table 2 Key distinguishing features between six *Ctenotus* species in the NSW Murray catchment.

Species		Black vertebral stripe	Pale spots	Page no.
Brown-blazed wedgesnout skink		Absent (tan band only)	On back and sides	106
Murray striped skink		Extends to hind legs only	Absent	108
Spotted-back skink		Extends onto tail	On back and sides	110
Royal skink		Extends to hind legs only	On sides only	112
Large striped skink		Extends onto tail	On sides only	114
Copper-tailed skink		Extends onto tail	Absent	116

rock skink potentially co-occurs with the other *Liopholis* species. The Snowy Mountain's rock skink and White's skink are not known to co-occur in the Murray catchment. All three species will retreat within a burrow system when disturbed, but will often reappear within a few minutes.

Eulamprus species (medium-bodied water skinks)

There are three *Eulamprus* species in the Murray catchment. These species can be distinguished by the presence of vertebral and dorsolateral stripes and the colour at the front of the ear opening. The southern water skink co-occurs with the other *Eulamprus* species. The yellow-bellied water skink and alpine water skink are not

Table 3 Key distinguishing features between three *Egernia* species in the NSW Murray catchment.

Species		Scale texture	Side colour and markings	Page no.
Cunningham's skink		Very rough with spiny tail	Dark with pale blotches	120
Black rock skink		Rough with raised scale ridges	Uniformly dark	122
Tree crevice skink		Rough with slight scale ridges	Dark with pale flecks	124

known to co-occur in the Murray catchment. All three species will take cover when disturbed, but will often regain their basking spot within a few minutes.

Pseudemoia species (High Country skinks)

There are four *Pseudemoia* species in the Murray catchment. These species can be distinguished by the number of dorsal stripes and the position of the dorsolateral

Table 4 Key distinguishing features between three *Liopholis* species in the NSW Murray catchment.

Species		Dorsal pattern	Side colour and markings	Page no.
Snowy Mountains rock skink		Pale stripes and pale spots	Dark band enclosing white spots	144
Tan-backed rock skink		Plain with no stripes or spots	Dark band with no white spots	146
White's skink		Plain or striped with spots	Pale with black-edged white spots	148

Table 5 Key distinguishing features between three *Eulamprus* species in the NSW Murray catchment.

Species		Body stripes	Colour of front margin of ear	Page no.
Yellow-bellied water skink		Absent	Black	126
Alpine water skink		Present	White	128
Southern water skink		Absent	White	130

stripe. The woodland tussock skink and Spencer's skink co-occur. The other two species have specialised habitat requirements and are unlikely to co-occur with any other species of *Pseudemoia*. The southern grass skink is found in treeless

Table 6 Key distinguishing features between four *Pseudemoia* species in the NSW Murray catchment.

Species		Dorsal markings	Dorsolateral stripe position	Page no.
Woodland tussock skink		1 dark vertebral stripe	Broken and positioned on 4th scale row from the centre of back	158
Southern grass skink		1-3 dark stripes	Continuous and positioned on 4th scale row from the centre of back	160
Glossy grass skink		1-3 dark stripes	Continuous and positioned on 3rd scale row from the centre of back	162
Spencer's skink		Single pale stripe and flecks	Thin and cream coloured	164

grasslands and the glossy grass skink is found in swamps and alpine bogs. With the exception of the rock-dwelling Spencer's skink, the other three species will seek the shelter of grass tussocks when disturbed.

Appendix 1 Reptiles of the Murray catchment and their associated vegetation types

(Key to vegetation symbols: AG = alpine grassland; AW = alpine woodland; MF = montane morest; WF = wet forest; DF = dry forest; GW = grassy box-gum woodland; SW = sandhill woodland; BB = black box woodland; RG = river red gum forest; BW = boree woodland; CS = chenopod shrubland; MW = mallee woodland; WL = wetlands).

Species name	AG	AW	MF	WF	DF	GW	SW	BB	RG	BW	CS	MW	WL
Broad-shelled turtle *Chelodina expansa*													
Long-necked turtle *Chelodina longicollis*													
Murray turtle *Emydura macquarii* ssp. *macquarii*													
Southern marbled gecko *Christinus marmoratus*													
Tessellated gecko *Diplodactylus tessellatus*													
Eastern stone gecko *Diplodactylus vittatus*													
Tree dtella *Gehyra variegata*													
Gibber gecko *Lucasium byrnei*													
Beaded gecko *Lucasium damaeum*													
Beaked gecko *Rhynchoedura ornata*													
Eastern spiny-tailed gecko *Strophurus intermedius*													
Thick-tailed gecko *Underwoodisaurus milii*													
Pink-tailed worm-lizard *Aprasia parapulchella*													
Olive legless lizard *Delma inornata*													
Burton's snake-lizard *Lialis burtonis*													
Eastern hooded scaly-foot *Pygopus schraderi*													
Sand goanna *Varanus gouldii* ssp. *gouldii*													
Heath monitor *Varanus rosenbergi*													
Lace monitor *Varanus varius*													
Jacky lizard *Amphibolurus muricatus*													
Nobbi dragon *Amphibolurus nobbi* ssp. *nobbi*													
Gippsland water dragon *Physignathus lesueurii* ssp. *howitti*													
Eastern bearded dragon *Pogona barbata*													
Mountain dragon *Rankinia diemensis*													
Eastern three-lined skink *Acritoscincus duperreyi*													
Red-throated skink *Acritoscincus platynotum*													
McCoy's skink *Anepischtos maccoyi*													
Southern rainbow skink *Carlia tetradactyla*													
Ragged snake-eyed skink *Cryptoblepharus pannosus*													

Appendix 1 Reptiles of the Murray catchment and their associated vegetation types

Species name	AG	AW	MF	WF	DF	GW	SW	BB	RG	BW	CS	MW	WL
Brown-blazed wedgesnout skink *Ctenotus allotropis*													
Murray striped skink *Ctenotus brachyonyx*													
Spotted-back skink *Ctenotus orientalis*													
Royal skink *Ctenotus regius*													
Large striped skink *Ctenotus robustus*													
Copper-tailed skink *Ctenotus taeniolatus*													
Alpine she-oak skink *Cyclodomorphus praealtus*													
Cunningham's skink *Egernia cunninghami*													
Black rock skink *Egernia saxatilis* ssp. *intermedia*													
Tree crevice skink *Egernia striolata*													
Yellow-bellied water skink *Eulamprus heatwolei*													
Alpine water skink *Eulamprus kosciusko*													
Southern water skink *Eulamprus tympanum*													
Three-toed skink *Hemiergis decresiensis* ssp. *talbingoensis*													
Delicate skink *Lampropholis delicata*													
Garden skink *Lampropholis guichenoti*													
Bougainville's skink *Lerista bougainvillii*													
Eastern robust slider *Lerista punctatovittata*													
Wood-mulch slider *Lerista timida*													
Snowy Mountains rock skink *Liopholis guthega*													
Tan-backed rock skink *Liopholis montana*													
White's skink *Liopholis whitii*													
Grey's skink *Menetia greyii*													
Saltbush skink *Morethia adelaidensis*													
Boulenger's skink *Morethia boulengeri*													
Coventry's skink *Niveoscincus coventryi*													
Woodland tussock skink *Pseudemoia entrecasteauxii*													
Southern grass skink *Pseudemoia pagenstecheri*													
Glossy grass skink *Pseudemoia rawlinsoni*													

Species name	AG	AW	MF	WF	DF	GW	SW	BB	RG	BW	CS	MW	WL
Spencer's skink *Pseudemoia spenceri*													
Blotched blue-tongue *Tiliqua nigrolutea*													
Shingleback *Tiliqua rugosa* ssp. *aspera*													
Eastern blue-tongue *Tiliqua scincoides* ssp. *scincoides*													
Dark-spined blind snake *Ramphotyphlops bicolor*													
Prong-snouted blind snake *Ramphotyphlops bituberculatus*													
Blackish blind snake *Ramphotyphlops nigrescens*													
Woodland blind snake *Ramphotyphlops proximus*													
Inland carpet python *Morelia spilota* ssp. *metcalfei*													
Common death adder *Acanthophis antarticus*													
Highlands copperhead *Austrelaps ramsayi*													
Eastern small-eyed snake *Cryptophis nigrescens*													
Yellow-faced whipsnake *Demansia psammophis* ssp. *psammophis*													
White-lipped snake *Drysdalia coronoides*													
Red-naped snake *Furina diadema*													
Eastern tiger snake *Notechis scutatus*													
Dwyer's snake *Parasuta dwyeri*													
Red-bellied black snake *Pseudechis porphyriacus*													
Western brown snake *Pseudonaja nuchalis*													
Eastern brown snake *Pseudonaja textilis*													
Curl snake *Suta suta*													
Eastern bandy bandy *Vermicella annulata*													

Appendix 2 Common and scientific names of other animals and plants mentioned

Amphibians
Salamander *Ambystoma* spp.
Turtles
Pig-nosed turtle *Carettochelys insculpta*
Legless lizards
Striped legless lizard *Delma impar*
Dragons
Canegrass dragon *Diporiphora winnecki*
Thorny devil *Moloch horridus*
Skinks
Alpine bog skink *Pseudemoia cryodroma*
Obscure skink *Morethia obscura*
Weasel skink *Saproscincus mustelinus*
Monitors
Komodo dragon *Varanus komodoensis*
Perentie *Varanus giganteus*
Short-tailed pygmy monitor *Varanus brevicauda*
Water monitor *Varanus salvator*
Snakes
Australian scrub python *Morelia kinghorni*
African rock python *Python sebae*
Asian rock python *Python molurus*
Blind snakes *Ramphotyphlops* spp.
Brown snakes *Pseudonaja* spp.
Carpet python *Morelia spilota* ssp.
Death adders *Acanthophis* spp.
Pygmy python *Antaresia perthensis*
Little whip snake *Parasuta flagellum*
Mulga snake *Pseudechis australis*

Reticulated python *Python reculatus*
Taipan *Oxyuranus* spp.
Tiger snake *Notechis scutatus*
White-bellied mangrove snake *Fordonia leucobalia*
Mammals
Black rat *Rattus rattus*
Cat *Felis catus*
European rabbit *Oryctolagus cuniculus*
Fat-tailed dunnart *Sminthopsis crassicaudata*
Red fox *Vulpes vulpes*
Yellow-footed antechinus *Antechinus flavipes*
Birds
Black kite *Milvus migrans*
Brown falcon *Falco berigora*
Bush stone-curlew *Burhinus grallarius*
Laughing kookaburra *Dacelo novaeguineae*
Pied butcherbird *Cracticus nigrogularis*
White-bellied sea-eagle *Haliaeetus leucogaster*
Overstorey plants
Alpine ash *Eucalyptus delegatensis*
Apple box *Eucalyptus bridgesiana*
Belah *Casuarina pauper*
Black box *Eucalyptus largiflorens*
Black cypress-pine *Callitris endlicheri*
Black sallee *Eucalyptus stellulata*
Blakely's red gum *Eucalyptus blakelyi*
Blackwood *Acacia melanoxylon*
Brittle gum *Eucalyptus mannifera*
Broad-leaved peppermint *Eucalyptus dives*
Buloke *Allocasuarina leuhmanii*
Currawang *Acacia doratoxylon*
Dropping sheoak *Allocasuarina verticillata*
Eurabbie *Eucalyptus bicostata*
Forest wattle *Acacia frigescens*
Grey box *Eucalyptus microcarpa*
Kurrajong *Brachychiton populeneus*
Long-leaved box *Eucalyptus goniocalyx*
Mountain ash *Eucalyptus regnans*
Mountain gum *Eucalyptus dalrympleana*
Mountain swamp gum *Eucalyptus camphora*
Murray pine *Callitris gracilis*
Narrow-leaved peppermint *Eucalyptus radiata*
Pointed mallee *Eucalyptus socialis*
Red box *Eucalyptus polyanthemos*
Red mallee *Eucalyptus oleosa*
Red stringybark *Eucalyptus macrorhyncha*

Ribbon gum *Eucalyptus viminalis*
Ridge-fruited mallee *Eucalyptus costata*
River red gum *Eucalyptus camaldulensis*
Snow gum *Eucalyptus pauciflora*
Swamp gum *Eucalyptus ovata*
Weeping pittosporum *Pittosporum angustifolium*
White box *Eucalyptus albens*
White cypress pine *Callitris glaucophylla*
White gum *Eucalyptus rossii*
Yellow Box *Eucalyptus melliodora*
Understorey plants
Blackthorn *Bursaria spinosa*
Boree *Acacia pendula*
Broad-leaved bitter-pea *Daviesia latifolia*
Butter bush *Pittosporum phylliraeoides*
Comb spider flower *Grevillea huegelii*
Cooba *Acacia salicina*
Daphne heath *Brachyloma daphnoides*
Deane's wattle *Acacia deanei*
Dogwood *Cassinia aculeata*
Dwarf cherry *Exocarpus strictus*
Emu bush *Eremophila longiflolia*
Gold-dust wattle *Acacia acinacea*
Golden wattle *Acacia pycnantha*
Grey mistletoe *Amyema quandang*
Grey wattle *Acacia brachybotrya*
Hazel pomaderris *Pomaderris aspera*
Hop bush *Dodonea viscose*
Kangaroo thorn *Acacia paradoxa*
Leafless cherry *Exocarpus aphyllus*
Leafy bossiaea *Bossiaea foliosa*
Narrow-leaf bitter-pea *Daviesia mimosoides*
Needlewood *Hakea leucoptera*
Miljee *Acacia oswaldii*
Mountain mirbelia *Mirbelia oxyloboides*
Montane wattle *Acacia montana*
Musk daisy-bush *Olearia argophylla*
Rosewood *Alectryon oleifolius*
River cooba *Acacia stenophylla*
Showy daisy-bush *Olearia pimelioides*
Silver wattle *Acacia dealbata*
Sweet quandong *Santalum acuminatum*
Water bush *Myoporium montanum*
Wedge-leaf hop-bush *Dodonaea cuneata*
Western silver wattle *Acacia decora*
Yarran *Acacia omalophylla*

Ground cover plants
Australian bindweed *Convolvulus erubescens*
Bidgee-widgee *Acaena novae-zelandiae*
Bladder saltbush *Atriplex vesicaria*
Black cotton-bush *Maireana decalvans*
Black bluebush *Maireana pyramidata*
Black roly-poly *Sclerolaena muricata*
Blown grass *Lachnagrostis filiformis*
Blue flax-lily *Dianella tasmanica*
Cactus pea *Bossiaea walkeri*
Cane grass *Eragrostis australasica*
Climbing saltbush *Einadia nutans*
Curly windmill grass *Enteropogon acicularis*
Common reed *Phragmites australis*
Common sneezeweed *Centipeda cunninghamii*
Cotton bush *Maireana aphylla*
Creeping saltbush *Atriplex semibaccata*
Cumbungi *Typha australis*
Dillon bush *Nitraria billardierei*
Fairy grass *Sporobolus caroli*
Giant rush *Juncus ingens*
Grey guinea-flower *Hibbertia obtusifolia*
Handsome flat-pea *Platylobium formosum*
Hard-head daisy *Brachyscome lineariloba*
Heaths *Epacris* spp.
Hop bitter-pea *Daviesia latifolia*
Kangaroo grass *Themeda triandra*
Knob sedge *Carex inverse*
Lesser joyweed *Alternanthera denticulate*
Lignum *Muehlenbeckia florulenta*
Mother shield fern *Polystichum proliferum*
Native carrot *Daucus glochidiatus*
Necklace fern *Asplenium flabellifolium*
Nitre goosefoot *Chenopodium nitrariaceum*
Nodding blue-lily *Stypandra glauca*
Old man saltbush *Atriplex nummularia*
Pale beauty-heads *Calocephalus sonderi*
Pale twinleaf *Zygophyllum glaucum*
Panic grass *Panicum* spp.
Plains grass *Austrostipa aristiglumis*
Purple burr-daisy *Calotis cuneifolia*
Quena *Solanum esuriale*
Ridge sida *Sida cunninghamii*
River bluebell *Wahlenbergia fluminalis*
River tussock *Poa labillardieri*
Rock fern *Cheilanthes sieberi*

Rough burr-daisy *Calotis scabiosifolia*
Rough spear grass *Austrostipa scabra*
Slender-fruit saltbush *Atriplex leptocarpa*
Small-flowered wallaby grass *Austrodanthonia setacea*
Small purslane *Calandrinia eremaea*
Spear grass *Austrostipa* spp.
Sphagnum moss *Sphagnum cristatum*
Spinifex (Porcupine grass) *Triodia scariosa*
Spiny rush *Eleocharis sphaceolatus*
Swamp dock *rumex brownii*
Thorny saltbush *Rhagodia spinescens*
Tussock grass *Poa* spp.
Tussock sedge *Carex appressa*
Wallaby grass *Austrodanthonia* spp.
Water couch *Paspalidium* spp.
Wattle mat-rush *Lomandra filiformis*
Wheat grass *Elymus scaber*
White-top wallaby grass *Austrodanthonia caespitosa*
Windmill grass *Chloris truncate*
Wire grass *Aristida ramosa*
Wire grass *Aristida behriana*

References

Reptile studies in the Murray catchment

Cunningham RB, Lindenmayer DB, Crane M, Michael DR and MacGregor CI (2007) Reptile and arboreal marsupial response to replanted vegetation in agricultural landscapes. *Ecological Applications* **17**: 609–619.

Daly G (2004) Surveys of reptiles and amphibians on the south-west slopes of New South Wales. *Herpetofauna* **34** (1): 2–6.

DLWC (2001) New South Wales Murray Catchment Pre-European Broad Vegetation Types. Department of Land and Water Conservation, Canberra, Australia.

Herring M (2003) 'Biodiversity in the billabong: a wildlife survey of the Eastern Billabong Creek catchment'. Unpublished report for *Heartlands*: CSIRO, Murray-Darling Basin Commission and the Eastern Billabong Landcare Groups.

Herring M (2005) 'Murrakool wildlife: magnificent diversity, precarious future'. Murrakool Inc.

Herring M, McGregor H, Herring J, Webb D and Knight A (2008) 'West Corurgan wildlife'. Murray Catchment Management Authority and Murray Wildlife Pty Ltd.

Herring M, McGregor H, Herring J, Webb D and Knight A (2008) 'Central Riverina wildlife'. Murray Catchment Management Authority and Murray Wildlife Pty Ltd.

Herring M, McGregor H, Webb D and O'Donnell D (2008) 'Restoration of the Jimaringle and Cockran Creek system'. Murray Wildlife Pty Ltd and Murray Irrigation Limited.

Herring M, Webb D and Pisasale M (2007) 'Murray wildlife – Murray land & water management plan wildlife survey 2005–2006'. Murray Wildlife Pty Ltd and Murray Irrigation Limited.

Lemckert F (1998) A survey of threatened herpetofauna of the south-west slopes of New South Wales. *Australian Zoologist* **30**: 492–500.

Michael DR (2004) Distribution, habitat preferences and conservation status of reptiles in the Albury-Wodonga region. *The Victorian Naturalist* **121** (5): 180–193.

Michael DR (2005) A record of the exotic Asiatic house gecko *Hemidactylus frenatus* being transported to Albury, New South Wales. *Herpetofauna* **35** (2): 101–102.

Michael DR (2008) A range extension for the yellow-faced whip snake *Demansia psammophis* in south-eastern NSW and an addition to the herpetofauna of the Albury-Wodonga region. *Herpetofauna* **37** (2): 81–82.

Michael DR, Cunningham RB and Lindenmayer DB (2008) A forgotten habitat? Granite inselbergs conserve reptile diversity in fragmented agricultural landscapes. *Journal of Applied Ecology* **45**: 1742–1752.

Michael DR, Cunningham RB and Lindenmayer DB (2009) Microhabitat relationships among five lizard species associated with granite outcrops in fragmented agricultural landscapes of south-eastern Australia. *Austral Ecology* DOI: 10.1111/j.1442-9993.2009.02029.x

Michael DR, Cunningham RB and Lindenmayer DB (2010) The social elite? Habitat heterogeneity, complexity and quality in granite inselbergs influence patterns of aggregation in *Egernia striolata* (Scincidae: Lygosominae). *Austral Ecology* (in press)

Michael DR and Herring MW (2005) Habitat of the pink-tailed worm-lizard *Aprasia parapulchella* in Albury, NSW. *Herpetofauna* **35** (2): 103–111.

Michael DR and Lindenmayer DB (2008) Records of the inland carpet python, *Morelia spilota metcalfei* (Serpentes: Pythonidae), from the south-western slopes of New South Wales. *Proceedings of the Linnean Society of New South Wales* **129**: 253–261.

Michael DR, Lindenmayer DB and Cunningham RB (2010) Managing rock outcrops to improve biodiversity conservation in Australian agricultural landscapes. *Ecological Management and Restoration* (in press).

Michael DR, Lindenmayer DB, Crane M, Montague-Drake R and McBurney L (2010) Reptilia, Murray catchment, New South Wales, South-eastern Australia. Checklist. (in press).

Michael DR, Crane M, Holliday S, Lindenmayer DB, MacGregor C, McBurney L and Montague-Drake R (2009) A range extension for the prong-snouted blindsnake *Ramphotyphlops bituberculatus* (Typhlopidae) in the South-western Slopes of NSW. *Herpetofauna* (in press).

Sass S, Watson D and Herring M (2007) A range extension for the blotched blue-tongue skink (*Tiliqua nigrolutea*) (Scincidae) and implications for its future survival. *Herpetofauna* **37** (2): 85–89.

Sass S, Watson D and Wilson A (2008) The reptile fauna of the upper Billabong Creek catchment area, southern New South Wales. *Herpetofauna* **38** (1): 41–50.

Spencer RJ (2006) The Murray River turtle, *Emydura macquarii*: population dynamics, nesting ecology and impact of the introduced red fox, *Vulpes vulpes*. Unpublished PhD thesis. University of Sydney.

Webster R, Belcher C and Leslie D (2003) A survey of threatened fauna in south-western New South Wales. *Australian Zoologist* **32**: 214–228.

Bibliography and further reading

Bennett A, Brown G, Lumsden L, Hespe D, Krasna S and Silins J (1998) Fragments for the future: wildlife in the Victorian Riverina (the northern plains). Department of Natural Resources and Environment, East Melbourne.

Bennett R (1997) *Reptiles and Frogs of the Australian Capital Territory.* National Parks Association of the ACT, Wodan, ACT.

Bull CM, Cooper SJB, Ben C and Baghurst BC (1998) Social monogamy and extra-pair fertilization in an Australian lizard, *Tiliqua rugosa. Behavioural Ecology and Sociobiology* **44**: 63–72.

Cogger HG (2000) *Reptiles and Amphibians of Australia.* 6th edn. Reed New Holland, Sydney.

Costermans L (1983) *Native Trees and Shrubs of South-eastern Australia*, Lansdowne Publishing, Sydney.

Coventry AJ and Robertson P (1991) *The Snakes of Victoria: A Guide to Their Identification.* Department of Conservation and Environment, Melbourne.

Doody JS, Guarino E, Georges A, Corey B, Murray G and Ewert MW (2006) Nest site choice compensates for climate effects on sex ratios in a lizard with environmental sex determination. *Evolutionary Ecology* **20**: 307–330.

Driver M and Porteners M (1993) *The use of locally native trees and shrubs in the southern Riverina*. Royal Botanic Gardens, Sydney.

Elphick M, Thomas J and Shine R (2006) Courtship and copulation in the southern water skink *Eulamprus heatwolei. Herpetofauna* **36** (1): 25–26.

Fischer J, Lindenmayer DB and Cowling A (2004) The challenge of managing multiple species at multiple scales: reptiles in an Australian grazing landscape. *Journal of Applied Ecology* **41**: 32–44.

Fry BG, Vidal N, Norman JA, Vonk FJ, Scheib H, Ramjan SFR, Kuruppu S, Fung K, Hedges SB, Richardson MK, Hodgson WC, Ignjatovic V, Summerhayes R and Kochva E (2005) Early evolution of the venom system in lizards and snakes. *Nature* **439**: 584–588.

Gardner MG, Hugall AF, Donnellan SC, Hutchinson MN and Foster R (2008) Molecular systematics of the social skinks: phylogeny and taxonomy of the *Egernia* group (Reptilia: Scincidae). *Zoological Journal of the Linnean Society* **154**: 781–794.

Greenlees M, Webb JK and Shine R (2005) Led by the blind: bandy bandy snakes *Vermicella annulata* (Elapidae) follow blindsnake chemical trails. *Copeia* **2005**: 184–187.

Greer AE (1989) *The Biology and Evolution of Australian Lizards.* Surrey Beatty and Sons, Chipping Norton, NSW.

Greer AE (1997) *The Biology and Evolution of Australian Snakes.* Surrey Beatty and Sons, Chipping Norton, NSW.

Heatwole HF and Taylor J (1987) *Ecology of Reptiles.* Surrey Beatty and Sons, Chipping Norton, NSW.

Henle K (1988) Population ecology and life history of the diurnal skink *Morethia boulengeri* in arid Australia. *Oecologia* **78**: 521–532.

Horner P (2007) Systematics of the snake-eyed skinks, *Cryptoblepharus* Wiegmann (Reptilia: Squamata: Scincidae) – an Australian-based review. *The Beagle, Records of the Museums and Art Galleries of the Northern Territory* **3**: 21–198.

Hutchinson M (2008) Nomenclature of the scincid lizards belonging to eastern Australian populations formerly assigned to *Lerista muelleri* (Fisher, 1881). *Herpetofauna* **38** (1): 39–40.

Jenkins R and Bartell R (1980) *Reptiles of the High Country.* Inkata Press, Melbourne.

Keith D (2004) Oceans shores to desert dunes: the native vegetation of New South Wales and the ACT. Department of Conservation and Environment, Hurstville, NSW.

Kent K, Earl G, Lunt I, Mullins B and Webster R (2002) Native Vegetation Guide for the Riverina Charles Sturt University, Berrigan Shire and Native Dog Landcare Group.

Koenig J, Shine R and Shea G (2001). The ecology of an Australian reptile icon: how do bluetongue lizards (*Tiliqua scincoides*) survive in suburbia? *Wildlife Research* **28**: 215–227.

Langkilde T, O'Connor D and Shine R (2003) Shelter-site use by five species of montane scincid lizards in south-eastern Australia. *Australian Journal of Zoology* **51**: 175–186.

Lindenmayer D, Claridge A, Hazell D, Michael D, Crane M, MacGregor C and Cunningham R (2003) *Wildlife on Farms: How to Conserve Native Animals.* CSIRO Publishing, Collingwood, Melbourne.

Lindenmayer DB, Crane M and Michael DR (2005) *Woodlands: A Disappearing Landscape.* CSIRO Publishing, Collingwood, Melbourne.

Masters P (1996) The effects of fire-driven succession on reptiles in spinifex grasslands at Uluru National Park, Northern Territory. *Wildlife Research* **23**: 39–47.

Michael DR, Lunt ID and Robinson WA (2003) Terrestrial vertebrate fauna of grasslands and grassy woodlands in Terrick Terrick National Park, northern Victoria. *Victorian Naturalist* **12** (5): 164–170.

Michael DR, Lunt ID and Robinson WA (2004) Enhancing fauna habitat in grazed native grasslands and woodlands: use of artificially placed log refuges by fauna. *Wildlife Research* **31**: 65–71.

Morrison SF, Keogh SJ and Scott IAW (2002) Molecular determination of paternity in a natural population of the multiply mating polygynous lizard *Eulamprus heatwolei*. *Molecular Ecology* **11**: 535–545.

O'Conner DE and Shine R (2006) Kin discrimination in the social lizard *Egernia saxatilis* (Scincidae). *Behavioral Ecology* **17**: 206–211.

Pianka ER (1969) Sympatry of desert lizards (*Ctenotus*) in Western Australia. *Ecology* **50**: 1012–1030.

Qualls CP and Shine R (1998) *Lerista bougainvillii*, a case study for the evolution of viviparity in reptiles. *Journal of Evolutionary Biology* **11**: 63–78.

Radder RS and Shine R (2007) Why do female lizards lay their eggs in communal nests? *Journal of Animal Ecology* **76**: 881–887.

Read JL (1995) Subhabitat variability: a key to the high reptile diversity in chenopod shrublands. *Austral Ecology* **20**: 494–501.

Read JL (1998) The ecology of sympatric scincid lizards (*Ctenotus*) in arid South Australia. *Australian Journal of Zoology* **46**: 617–629.

Read JL and Bowen Z (2001) Population dynamics, diet and aspects of the biology of feral cats and foxes in arid South Australia. *Wildlife Research* **28**: 195–203.

Shine R (1991) *Australian Snakes: A Natural History*. Reed New Holland, Sydney.

Shine R (1994) 'The biology and management of diamond pythons (*Morelia s. spilota*) and carpet pythons (*M. s. variegata*)'. Species Management Report Number 15, New South Wales National Parks and Wildlife Service.

Shine R (1999) Why is sex determined by nest temperature in many reptiles? *Trends in Ecology and Evolution* **14**: 186–189.

Shine R (1999) Egg-laying reptiles in cold climates: determinants and consequences of nest temperatures in montane lizards. *Journal of Evolutionary Biology* **12**: 918–926.

Shine R (2005) Life-history evolution in reptiles. *Annual Review of Ecology, Evolution and Systematics* **36**: 23–46.

Spencer R and Thompson MB (2005) Experimental analysis of the impact of foxes on freshwater turtle populations, *Conservation Biology* **19**: 845–854.

Stelling F (1998) 'South West Slopes revegetation guide: south of the Murrumbidgee River'. Murray Catchment Management Committee & Department of Land and Water Conservation, Albury, NSW.

Stow AJ and Sunnucks P (2004) Inbreeding avoidance in Cunningham's skinks (*Egernia cunninghami*) in natural and fragmented habitat. *Molecular Ecology* **13**: 443–447.

Swan G, Shea G and Sadlier R (2004) *A Field Guide to the Reptiles of New South Wales*. 2nd edn. New Holland Publishers, Sydney.

Swan G and Wilson S (2008) *What Snake Is That? Introducing Australian Snakes*. Reed New Holland, Sydney.

Swanson S (2007) *Field Guide to Australian Reptiles*. Steve Parish Publishing, Acherfield, Queensland.

Tzaros C (2005) *Wildlife of the Box-Ironbark Country*. CSIRO Publishing, Collingwood, Melbourne.

Whitaker P, Ellis K and Shine R (2000) The defensive strike of the eastern brownsnake (*Pseudonaja textilis*, Elapidae). *Functional Ecology* **14**: 25–31.

Wilson S and Swan G (2008) *A Complete Guide to Reptiles of Australia*. 2nd edn. New Holland Publishers, Sydney.

Index - scientific names

Index - common names

www.ingramcontent.com/pod-product-compliance
Lightning Source LLC
LaVergne TN
LVHW072012300926
849160LV00002B/8